Sigrun Beige

Long-term and Mid-term Mobility Decisions during the Life Course

Sigrun Beige

Long-term and Mid-term Mobility Decisions during the Life Course

Findings from a retrospective survey

Südwestdeutscher Verlag für Hochschulschriften

Impressum/Imprint (nur für Deutschland/ only for Germany)
Bibliografische Information der Deutschen Nationalbibliothek: Die Deutsche Nationalbibliothek verzeichnet diese Publikation in der Deutschen Nationalbibliografie; detaillierte bibliografische Daten sind im Internet über http://dnb.d-nb.de abrufbar.
 Alle in diesem Buch genannten Marken und Produktnamen unterliegen warenzeichen-, marken- oder patentrechtlichem Schutz bzw. sind Warenzeichen oder eingetragene Warenzeichen der jeweiligen Inhaber. Die Wiedergabe von Marken, Produktnamen, Gebrauchsnamen, Handelsnamen, Warenbezeichnungen u.s.w. in diesem Werk berechtigt auch ohne besondere Kennzeichnung nicht zu der Annahme, dass solche Namen im Sinne der Warenzeichen- und Markenschutzgesetzgebung als frei zu betrachten wären und daher von jedermann benutzt werden dürften.

Verlag: Südwestdeutscher Verlag für Hochschulschriften Aktiengesellschaft & Co. KG
Dudweiler Landstr. 99, 66123 Saarbrücken, Deutschland
Telefon +49 681 37 20 271-1, Telefax +49 681 37 20 271-0
Email: info@svh-verlag.de
Zugl.: Zürich, ETH Zürich, Dissertation, 2008

Herstellung in Deutschland:
Schaltungsdienst Lange o.H.G., Berlin
Books on Demand GmbH, Norderstedt
Reha GmbH, Saarbrücken
Amazon Distribution GmbH, Leipzig
ISBN: 978-3-8381-0139-2

Imprint (only for USA, GB)
Bibliographic information published by the Deutsche Nationalbibliothek: The Deutsche Nationalbibliothek lists this publication in the Deutsche Nationalbibliografie; detailed bibliographic data are available in the Internet at http://dnb.d-nb.de.
 Any brand names and product names mentioned in this book are subject to trademark, brand or patent protection and are trademarks or registered trademarks of their respective holders. The use of brand names, product names, common names, trade names, product descriptions etc. even without a particular marking in this works is in no way to be construed to mean that such names may be regarded as unrestricted in respect of trademark and brand protection legislation and could thus be used by anyone.

Publisher: Südwestdeutscher Verlag für Hochschulschriften Aktiengesellschaft & Co. KG
Dudweiler Landstr. 99, 66123 Saarbrücken, Germany
Phone +49 681 37 20 271-1, Fax +49 681 37 20 271-0
Email: info@svh-verlag.de

Printed in the U.S.A.
Printed in the U.K. by (see last page)
ISBN: 978-3-8381-0139-2

Copyright © 2010 by the author and Südwestdeutscher Verlag für Hochschulschriften Aktiengesellschaft & Co. KG and licensors
All rights reserved. Saarbrücken 2010

Table of content

1. Introduction 1
2. Long-term and mid-term mobility decisions 4
   - 2.1 General description 4
   - 2.2 Places of residence and moving behaviour 6
   - 2.3 Ownership of mobility tools 8
3. Life course 10
   - 3.1 Concept and structure of the life course 10
   - 3.2 Qualitative and quantitative approaches 12
   - 3.3 Description of the life course 13
     - 3.3.1 General description 14
     - 3.3.2 Places of residence and moving behaviour 16
     - 3.3.3 Ownership of mobility tools 19
4. Research questions 22
5. Methodology 24
   - 5.1 Discrete choice modelling 24
     - 5.1.1 The logit model 25
     - 5.1.2 The probit model 28
     - 5.1.3 Model estimation 28
   - 5.2 Duration modelling 29
6. Data 31
   - 6.1 Data requirements 31
   - 6.2 Data collection 31
     - 6.2.1 Questionnaire 33
     - 6.2.2 Sampling 34
     - 6.2.3 Description of the survey and response 35
   - 6.3 Representativeness of the data 37
     - 6.3.1 Representativeness of the household sample 38
     - 6.3.2 Representativeness of the person sample 41
     - 6.3.3 Comparison to the Swiss Household Panel 45

7	Analyses for the year 2005		51
	7.1	Spatial and transport system defined classification	51
	7.2	Description of the households	54
	7.3	Description of the persons	58
		7.3.1 General description	58
		7.3.2 Places of residence and moving behaviour	65
		7.3.3 Ownership of mobility tools	66
8	Analyses for the period from 1985 to 2004		86
	8.1	General description	86
	8.2	Places of residence and moving behaviour	88
	8.3	Ownership of mobility tools	98
	8.4	Duration modelling for long-term and mid-term mobility	116
	8.5	Changes in long-term and mid-term mobility	134
9	Summary of the analyses and conclusions		149
	9.1	Summary of the analyses	149
	9.2	Conclusions and implications for policy and planning	159
10	Outlook		164
11	References		166

Appendix

List of tables

Table 1	Household size by municipality	38
Table 2	Household type by municipality	39
Table 3	Gender of all household persons by municipality	40
Table 4	Age of all household persons by municipality	41
Table 5	Gender of all persons by municipality	42
Table 6	Age of all persons by municipality	43
Table 7	Gender, age and place of residence of all persons five years ago	44
Table 8	Household and accommodation description	46
Table 9	Person description	47
Table 10	Gender, age and residential mobility	48
Table 11	Variance analysis of the residential mobility	49
Table 12	Shares of households and persons by ARE classification of their place of residence (2005)	53
Table 13	Household size by ARE classification (2005)	54
Table 14	Household type by ARE classification (2005)	54
Table 15	Gender of all household persons by ARE classification (2005)	55
Table 16	Age of all household persons by ARE classification (2005)	55
Table 17	Household income per month by ARE classification (2005)	56
Table 18	Accommodation size, type and costs by ARE classification (2005)	57
Table 19	Household vehicles by ARE classification (2005)	58
Table 20	Gender of the persons by ARE classification (2005)	59
Table 21	Age of the persons by ARE classification (2005)	59
Table 22	Nationality of the persons by ARE classification (2005)	60

Table 23	Occupation of the persons by ARE classification (2005)	60
Table 24	Education of the persons by ARE classification (2005)	61
Table 25	Employment of the persons by ARE classification (2005)	61
Table 26	Mode of transport to education and employment by ARE classification only considering persons in education or employment (2005)	62
Table 27	Person income per month by ARE classification (2005)	63
Table 28	Satisfaction of the persons by ARE classification (2005)	64
Table 29	Residential mobility by ARE classification (2005)	65
Table 30	Reasons for the last move by ARE classification (2005)	66
Table 31	Probability for a move within the next year by ARE classification (2005)	66
Table 32	Driving licence ownership and car availability by ARE classification (2005)	67
Table 33	Public transport season ticket ownership by ARE classification (2005)	67
Table 34	Car availability and public transport season ticket ownership (2005)	68
Table 35	Description of the explanatory variables (2005)	70
Table 36	Binomial logit models for car availability and public transport season ticket ownership (2005)	72
Table 37	Binomial logit models for car availability and public transport season ticket ownership only considering persons in education or employment (2005)	74
Table 38	Mobility tool ownership in groups by ARE classification (2005)	75
Table 39	Binomial logit models for mobility tool ownership in groups (2005)	76
Table 40	Multinomial logit model for mobility tool ownership in groups (2005)	77
Table 41	Nested logit model for mobility tool ownership in groups with two nests for car and no car (2005)	79
Table 42	Cross-nested logit model for mobility tool ownership in groups with four nests for car, national and regional tickets, half-fare discount tickets and no mobility tools (2005)	81

Table 43	Multivariate probit model for mobility tool ownership in groups (2005)	83
Table 44	Description of the various models for mobility tool ownership in groups (2005)	85
Table 45	Residential durations by ARE classification (1985-2004)	89
Table 46	Residential distances by ARE classification (1985-2004)	90
Table 47	Directions of all moves by ARE classification (1985-2004)	91
Table 48	Directions of all moves by accommodation size (1985-2004)	92
Table 49	Reasons for all moves by ARE classification (1985-2004)	92
Table 50	Distance to the places of education by ARE classification (1985-2004)	95
Table 51	Distance to the places of employment by ARE classification (1985-2004)	95
Table 52	Mobility tool ownership by ARE classification (1985-2004)	99
Table 53	Description of the explanatory variables (1985-2004)	103
Table 54	Binomial logit models for car availability and public transport season ticket ownership (1985-2004)	106
Table 55	Mobility tool ownership in groups by ARE classification (1985-2004)	108
Table 56	Changes in mobility tool ownership in groups (1985-2004)	108
Table 57	Multinomial logit model for mobility tool ownership in groups (1985-2004)	109
Table 58	Nested logit model for mobility tool ownership in groups with two nests for car and no car (1985-2004)	111
Table 59	Cross-nested logit model for mobility tool ownership in groups with four nests for car, national and regional tickets, half-fare discount tickets and no mobility tools (1985-2004)	112
Table 60	Multivariate probit model for mobility tool ownership in groups (1985-2004)	114
Table 61	Description of the various models for mobility tool ownership in groups (1985-2004)	116
Table 62	Hazard ratios of the duration models for the residential, education and employment durations (1985-2004)	122

Table 63	Hazard ratios of the duration models for the car availability and public transport season ticket ownership durations (1985-2004)	124
Table 64	Hazard ratios of the competing risks duration models for the residential, education and employment durations (1985-2004)	127
Table 65	Hazard ratios of the competing risks duration models for the car availability and public transport season ticket ownership durations (1985-2004)	129
Table 66	Hazard ratios of the combined duration models for the different types of durations (1985-2004)	131
Table 67	Multinomial logit model for the different types of durations using the residential durations as referential category (1985-2004)	133
Table 68	Changes in the places of residence, education and employment within the same year (1985-2004)	134
Table 69	Correlations between all changes within the same year (1985-2004)	135
Table 70	Binomial logit models for the changes in residence, education and employment (1985-2004)	139
Table 71	Binomial logit models for the changes in car availability and public transport season ticket ownership (1985-2004)	140
Table 72	Hazard ratios of the duration models for the delays following a move until the next change in car availability and public transport season ticket ownership (1985-2004)	147
Table 73	Hazard ratios of the duration models for the delays following a change in education or employment until the next change in car availability and public transport season ticket ownership (1985-2004)	148

List of figures

Figure 1	Intra-regional and inter-regional migration	5
Figure 2	Temporal dimensions concerning the life course	11
Figure 3	Comparison of the structures in MNL, NL and CNL models	28
Figure 4	Location of the study areas	35
Figure 5	Response rate in regard to contact by telephone at municipal level	36
Figure 6	Response rate over time in regard to contact by telephone at municipal level	37
Figure 7	Gender, age and residential mobility	48
Figure 8	Residential mobility	50
Figure 9	Spatial and transport system defined classification (ARE) of all Swiss municipalities	52
Figure 10	Spatial and transport system defined classification (ARE) of the study areas	53
Figure 11	Satisfaction of the persons by ARE classification (2005)	64
Figure 12	Car availability and public transport season ticket ownership (2005)	69
Figure 13	Structure of the nested logit model for mobility tool ownership in groups with two nests for car and no car	79
Figure 14	Structure of the cross-nested logit model for mobility tool ownership in groups with four nests for car, national and regional tickets, half-fare discount tickets and no mobility tools	81
Figure 15	Personal and familial events by time (1985-2004)	87
Figure 16	Personal and familial events by age (1985-2004)	87
Figure 17	Distribution of the residential durations (1985-2004)	88
Figure 18	Distribution of the residential distances (1985-2004)	90
Figure 19	Persons in education and employment as well as the monthly income by time (1985-2004)	93

Figure 20	Persons in education and employment as well as the monthly income by age (1985-2004)	94
Figure 21	Median distances to the places of education and employment by time (1985-2004)	96
Figure 22	Median distances to the places of education and employment by age (1985-2004)	96
Figure 23	Mode of transport to education and employment by time (1985-2004)	97
Figure 24	Mode of transport to education and employment by age (1985-2004)	98
Figure 25	Mobility tool ownership by time (1985-2004)	100
Figure 26	Mobility tool ownership by age (1985-2004)	100
Figure 27	Car ownership by gender, age and birth cohort membership (1985-2004)	102
Figure 28	National and regional ticket ownership by gender, age and birth cohort membership (1985-2004)	102
Figure 29	Half-fare discount ticket ownership by gender, age and birth cohort membership (1985-2004)	103
Figure 30	Distribution of the residential, education and employment durations (1985-2004)	117
Figure 31	Distribution of the car availability and public transport season ticket ownership durations (1985-2004)	118
Figure 32	Hazard rates of the duration models for the residential, education and employment durations (1985-2004)	119
Figure 33	Hazard rates of the duration models for the car availability and public transport season ticket ownership durations (1985-2004)	120
Figure 34	Hazard rates of the competing risks duration models for the residential, education and employment durations (1985-2004)	125
Figure 35	Hazard rates of the competing risks duration models for the car availability and public transport season ticket ownership durations (1985-2004)	126
Figure 36	Changes in residence, education and employment by time (1985-2004)	136

Figure 37 Changes in residence, education and employment by age (1985-2004) 136

Figure 38 Changes in car availability and public transport season ticket ownership by time (1985-2004) 137

Figure 39 Changes in car availability and public transport season ticket ownership by age (1985-2004) 138

Figure 40 Distribution of the delays following a move until the next change in the places of education and employment, and vice versa (1985-2004) 142

Figure 41 Distribution of the delays following a move until the next change in car availability and public transport season ticket ownership (1985-2004) 143

Figure 42 Distribution of the delays following a change in education or employment until the next change in car availability and public transport season ticket ownership (1985-2004) 144

Figure 43 Hazard rates for the delays following a move until the next change in car availability and public transport season ticket ownership (1985-2004) 145

Figure 44 Hazard rates of the delays following a change in education or employment until the next change in car availability and public transport season ticket ownership (1985-2004) 146

Acknowledgements

First of all, I would like to thank my parents and my family, who gave me their support throughout my life.

Furthermore, I would like to express my sincere gratitude to my examiner Professor Kay W. Axhausen from the ETH Zürich, who inspired this thesis, shared his expertise and ideas as well as gave me useful advice and comments. My gratitude also goes to my two co-examiners Professor Rico Maggi from the Università della Svizzera Italiana in Lugano and Professor Michael Wegener from Spiekermann & Wegener Stadt- und Regionalforschung in Dortmund for their invaluable contributions to the thesis at hand. Moreover, I would like to thank Professor Darren M. Scott from the McMaster University in Hamilton for his valuable input concerning the modelling approaches and methods as well as for providing the opportunity to carry out research at the McMaster University.

Special thanks go to the municipalities of Bassersdorf, Bülach, Dietlikon, Dübendorf, Horgen, Kloten, Rümlang, Wallisellen, Wangen-Brüttisellen, Winterthur and Zürich for their support with respect to the retrospective survey concerning long-term and mid-term mobility decisions during the life course. In this context, I am especially grateful to all the households and persons that participated in this survey and, therefore, made this thesis possible in the first place.

In addition, I wish to acknowledge the help and important support from my colleagues during the time at the IVT, especially in the transport planning group.

And, last, but not least, I owe much to all the people giving me encouraging and cheering words when they were necessary as well as readjusting the work to a more realistic size when it seemed to become too enormous.

And Olaf, thank you for the music.

Life can only be understood backwards.
In the meantime, it has to be lived forwards.

Søren Kierkegaard

1 Introduction

Long-term and mid-term mobility of people involves on the one hand decisions about their residential locations and the corresponding moves. In this context, distance and direction, frequency of moves and durations of stays as well as reasons for moving are of central interest (Wagner, 1990). At the same time, the places of education and employment play an important role. On the other hand, the ownership of mobility tools, such as cars and different public transport season tickets, is a complementary element in this process, which also binds substantial resources. Overall, the ownership or non-ownership of mobility tools is relatively stable over longer periods of time (Axhausen and Beige, 2003; Simma and Axhausen, 2003). These two aspects of mobility behaviour are closely connected to one another.

At the same time, there exist strong interrelationships with daily travel behaviour. Short-term mobility is affected by the location of the places of residence, education and employment. Due to changes in these spatial structures, e.g., due to moves or changes in occupation, the availability as well as the quality and quantity of the available transport systems change. In turn, the decisions about the ownership and usage of the various mobility tools are influenced, as they provide access to the different transport systems and determine the marginal costs of usage. In this context, the question arises, to what extent the availability of mobility tools already affects the residential and occupational decisions of people, particularly with regard to so-called self-selection processes (Cao, Mokhtarian and Handy, 2006; Prillwitz, Harms and Lanzendorf, 2006). The relationship between migration and mobility becomes much closer here. In particular, it is suggested that the ability to commute over longer distances, without significantly increasing travel times, gives people the opportunity to substitute residential relocation by commuting. This leads to a reduction in residential mobility due to changes in occupation (Pooley, Turnbull and Adams, 2005).

Daily travel behaviour is characterised by habits and routines (Lanzendorf, 2006; Prillwitz and Lanzendorf, 2006). At the same time, decisions concerning long-term and mid-term mobility have lasting effects, since corresponding changes involve substantial amounts of resources (costs, time, etc.). Therefore, it is necessary to analyse their dynamics over longer periods of time.

A longitudinal perspective on the relationships between residential mobility and mobility tool ownership is available from people's life courses, which link different dimensions of life together. Besides personal and familial history, locations of residence, education and employment as well as the ownership of mobility tools can be taken into account. These life course dimensions are usually not independent from one another. Events in one area are frequently connected to changes in other areas. Decisions are rarely made in isolation and choice behaviour is often context dependent (Verhoeven, Arentze, Timmermans and van der Waerden, 2005). At the same time, this longitudinal approach provides the possibility to observe developments over time, as behaviour is influenced by time, and identify state dependencies (Hensher, 1998; Hollingworth and Miller, 1996; Lanzendorf, 2003; Verhoeven et al., 2005; Wagner, 1990). The life course perspective enables the integration of the temporal dimension and dynamics into the analyses of long-term and mid-term mobility in a comprehensive way (Elder, 2000). The structure of the life course is described with trajectories and transitions (Elder, 2000). As intermediary concept, the life course is seen as a sequence of events (Elder, 2000; Sackmann and Wingens, 2001). In this context, it is worthwhile to understand an event as well as the history leading up to its occurrence, since past behaviour is strongly correlated to present behaviour (Box-Steffensmeier and Jones, 2004). A further advantage regarding the investigation and improved understanding and modelling of the dynamic aspects concerning mobility is the provision of more accurate and coherent forecasts of the future (Lanzendorf, 2003).

Concerning the analysis of residential mobility, there is the benefit of taking into account resident and mobile people at the same time, since the respondents both stay and move during their life course (Wagner, 1990). Usually residential mobility studies overemphasise the event of moving and give relatively little attention to the periods of stability (Clark, Deurloo and Dieleman, 2003). In general, overall residential mobility arises from high mobility of a smaller sub-population, thus, the whole population can be divided into stayers and movers (Wagner, 1990).

Analysing people's life courses can contribute to the understanding of their reactions to changes occurring in their personal and familial life, within their household as well as in the spatial structures (Simma and Axhausen, 2003). For instance, one can analyse how a move affects mobility tool ownership and, therefore, travel behaviour, since mobility tool ownership can be used as a proxy for the actual travel behaviour (Prillwitz et al., 2006; Simma and Axhausen, 2003).

Changes during the life course play a central role, when formulating transport policies, which are designed to influence people's behaviour, as they reconsider and reflect their decisions and choices only in the cases where the situation is very different from the usual context (Gorr, 1997; Jones, Dix, Clarke and Heggie, 1983; Lanzendorf, 2003). Thereby, questions regarding how, when and why such changes might happen are of large interest for policy makers and planners. From the incorporation of temporal effects, besides spatial effects, into the analyses of long-term and mid-term mobility a better assessment of the impact of policies and other interventions on travel behaviour is expected (Lanzendorf, 2006).

These dynamic effects can not be captured with cross-sectional data (Dargay, 2001). The analyses of long-term and mid-term mobility decisions require corresponding longitudinal data that describes people's life courses. Solely this kind of data enables the investigation of continuity and change over time (Ryder, 1965). In order to collect such data, a retrospective survey covering the 20 year period from 1985 to 2004 was carried out at the beginning of 2005 in a stratified sample of municipalities in the Zurich region, Switzerland, taking into account different spatial and transport system defined municipality types (Beige and Axhausen, 2005).

The structure of the thesis at hand is as follows: After the introduction, the second chapter gives a description concerning long-term and mid-term mobility decisions, covering the aspect of residential mobility as well as the aspect of the ownership of mobility tools, while the third chapter considers these two aspects in the context of the life course approach in more detail. In the fourth chapter the research questions explored in the thesis are specified. Subsequently the methodology is illustrated. The methods applied, in order to address the research questions, are on the one hand univariate and multivariate discrete choice modelling as well as on the other hand duration modelling, also called event history modelling. In the sixth chapter the longitudinal data collected in the retrospective survey is described. The seventh and the eighth chapters concentrate on the analyses for the year 2005 and for the period from 1985 to 2004, respectively. The main focus lies on the two aspects of long-term and mid-term mobility, looking at the dynamics of mobility tool ownership and the relationships with residential choices as well as with locations of education and employment. In this context various discrete choice models are described and compared with one another. Furthermore, the durations, changes and delays between changes are examined, using event history modelling. Then, the results of the analyses are summarised with respect to the research questions stated in the fourth chapter, as well as corresponding conclusions are drawn and implications for policy and planning are specified. Finally, the outlook is presented.

2 Long-term and mid-term mobility decisions

2.1 General description

On the one hand, mobility describes an actual movement of people and their observable behaviour (Franz, 1984). On the other hand, mobility also stands for the ability to move. In this context, the term motility is proposed as an alternative. Motility refers to potential mobility at the individual level. It is defined, as the way in which a person appropriates what is possible in the area of mobility and uses this potential for his or her activities (Kaufmann, 2002). Motility is acquired, for instance, in form of mobility resources, and then transformed into mobility. The propensity to be mobile varies from one person to another (Kaufmann, 2002). Increasingly motility becomes a form of capital, e.g., by acquiring skills for and access to a number of systems. In this context, motility is not necessarily there to be transformed into mobility (Kaufmann, 2002). Usually, mobility is not carried out as a purpose in itself, but rather as a mean to reach a certain objective (Franz, 1984; Jones *et al.*, 1983). Mobility is the result of a decision process of individuals and their families (Franz, 1984).

In general, mobility refers to the change of an individual between the units of a defined system (Franz, 1984). Thereby, spatial mobility relates to the spatial system. One approach to spatial mobility differentiates between migration, residential mobility, travel and daily travel behaviour (Kaufmann, 2002). Each is linked to specific temporalities, life history for migration, the life course and the year for residential mobility, the year and the month for travel, the week and the day for daily travel behaviour. These different forms of mobility influence one another. Those linked to longer temporal durations, such as the life history and the life course, have a systematic impact on the shorter ones. After relocation, people inevitably show a travel behaviour that is different from the travel behaviour before relocation, as a move usually implies changes in the spatial distribution of activities and, hence, the activity space (Chapin, 1965; Chapin, 1974; Hägerstrand, 1970; Scheiner, 2006). The activity space of an individual consists of the different locations of activities frequented by this individual over a period of time (Schönfelder, 2006). The most important activities are housing, education, employment, shopping and leisure. Normally the place of residence is the central point of reference of an individual activity space, which is the location where most trips start and end (Franz, 1984). Residential mobility is differentiated by the extent the activity space changes after a move. This strongly depends on the distance of the move (Scheiner, 2006). In the case of intra-regional migration certain activity places continue to be frequented, whereas inter-regional migration leads to a change of all activity places, i.e., the entire previous activity space is abandoned and rebuilt at the new place of residence (Scheiner, 2006). Figure 1 shows the activity spaces for intra-regional and inter-regional migration (Franz, 1984). When occupation is considered in this context, inter-regional mobility takes place, when the location of occupation is changed. Changing the place of residence is, therefore, only a result of an occupational change. Regarding intra-regional mobility, this is not necessarily the case, as a move takes place, although the place of occupation remains the same (Franz, 1984).

Figure 1   Intra-regional and inter-regional migration

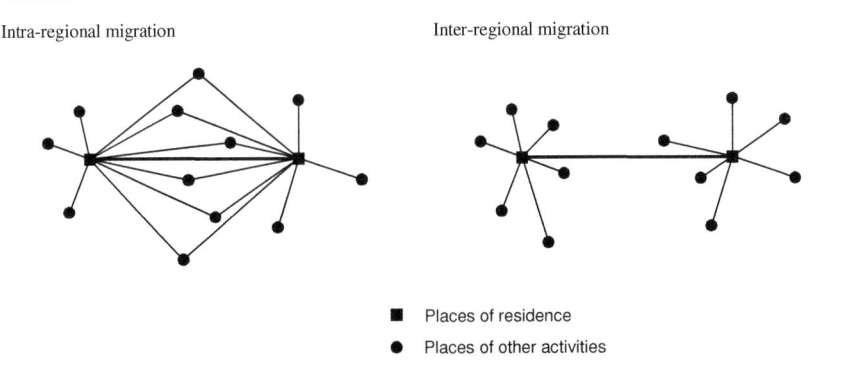

There exist mutual interrelationships between residential mobility and travel behaviour (Scheiner, 2006). However, the direction of cause and effect is not clear. Residential mobility influences travel behaviour in a number of ways. These influences are considered more frequently in research than the reverse impacts of travel behaviour on residential mobility, due to certain conceptual models of spatial mobility that propose a sequential, hierarchical process of long-term, mid-term and short-term decisions. Salomon and Ben-Akiva (1983) regard choices of residential and occupational locations as well as mobility tool ownership and the mode of commuting as long-term and mid-term decisions, while short-term decisions on other daily mobility, e.g., with respect to trip frequency, mode, destination, route and time of the day, are based on the longer term decisions. Activity spaces and, therefore, trip distances, trip timing and the use of the transport modes strongly correspond to the decisions on locations (Scheiner, 2006). At the same time, the transport systems play an additional role for locational decisions. The impact of individual travel behaviour on residential mobility is illustrated with regard to the effect of car availability which considerably increases the possibilities of households' residential choices, while the choices of households without a car depend on access to public transport and other modes. Furthermore, the locations and distances to daily activity places as well as the quality and quantity of spatial opportunities affect residential decisions (Scheiner, 2006). In this context, self-selection refers to the tendency of people to choose locations based on their travel preferences, abilities and behaviour (Cao *et al.*, 2006; Prillwitz *et al.*, 2006). Hence, spatial mobility is a process of intertwined long-term, mid-term and short-term decisions (Scheiner, 2006).

In general, spatial choices involve a combination of deciding on residential and occupational locations as well as on daily commuting (Kaufmann, 2002; Van Ommeren, Rietveld and Nijkamp, 1999; Van Ommeren, Rietveld and Nijkamp, 2000). Commuting is the consequence of a spatial discrepancy between the residential and occupational locations (Rouwendal and van der Vlist, 2005). In this context, commuting is a substitution for migration, in order to avoid relocation, implying a transformation of mobility temporalities from long-term and mid-term to short-term (Kaufmann, 2002). Commuting distance and time are changed either by residential mobility or occupational mobility. These two types of mobility are closely connected, especially in the sense that observing one type of mobility makes it more likely to observe the other one as well (Rouwendal and van der Vlist, 2005). Changing the place of occupation over a long distance necessitates a residential move, whereas for shorter residential moves, i.e., within a housing and labour market area, it is generally possible to choose the residential location without reference to the occupational location, at least, when the commuting distance is not too long (Dieleman, 2001). The decision concerning the question of migrating or commuting is primarily based on a trade-off between migration costs, i.e., the actual costs of moving, administration, etc., commuting costs, in terms of distance and time, as well as housing and labour market conditions (Van Ommeren *et al.*, 1999; Van Ommeren *et al.*, 2000).

2.2 Places of residence and moving behaviour

Residential mobility is the primary mean of making adjustments in housing consumption (Rossi, 1955).

Various variables significantly affect residential mobility. In the literature, age is most consistently reported, showing an inverse relationship to the number of moves (Clark and Onaka, 1983; Courgeau, 1985; Vandersmissen, Séguin, Thériault and Claramunt, 2005). A higher education and employment status is associated with more changes in residence (Courgeau, 1985). Changes in occupation also lead to a higher number of moves (Hollingworth and Miller, 1996). At the same time, residential mobility is less dependent on the absolute income and more dependent on variations in income. An increase in income encourages residential mobility, while a decrease in income seems to have no effect. The influence of the household structure is rather ambiguous (Vandersmissen et al., 2005). Housing characteristics also play an important role, such as type, size, space adequacy and the tenure status. Concerning space adequacy, defined as the ratio of the actual number and the required number of bedrooms, Hollingworth and Miller (1996) found that households lacking space are more mobile, while households with excess space are not. Generally, households gain space after a move, except for single-person households and older households (Vandersmissen et al., 2005). Renters are about twice as likely to move as owners, because the transaction costs of owning are substantially higher than those of renting (Hollingworth and Miller, 1996). In this context, residential mobility is closely related to the situation on the housing market and its conditions (Aufhauser, 1995; Clark et al., 2003; Dieleman, 2001). Accessibility to the places of occupation influences the residential mobility in such a way that with increasing travel distance the probability for moving also rises, but this is not a strong explanatory variable (Beige, 2006; Blijie, 2005). Preferences concerning the provision of transport facilities are varied, depending on the availability of mobility tools, especially a car (Blijie, 2005; Zondag and Pieters, 2005). However, accessibility considerations and transport related criteria are significantly less important than housing and neighbourhood related attributes (Zondag and Pieters, 2005).

Concerning the reasons for moving, Clark and Onaka (1983) distinguish first between involuntary and voluntary moves. Voluntary changes of residence are further divided into adjustment moves and induced moves. Thereby, adjustment moves are considerably more frequently expressed as reasons for moving than induced moves. Adjustment moves are intended to alter the quality and quantity of housing consumption. It is possible to classify these reasons into the following three categories: housing characteristics, neighbourhood characteristics, and accessibility. Among the housing characteristics, space is usually the dominant factor influencing the decision to move. Quality or design aspects considerations are less important. The change in tenure is especially important in the movement of households from rented to owned housing. Although tenure is not an inherent characteristic of housing, most households need to relocate, in order to change their tenure status. The influence of neighbourhood quality is of a relatively ambiguous and complex nature. Accessibility concerns the location of the places of education, employment, shopping and leisure as well as the accessibility to family and friends. Induced moves are usually related to changes in other areas of life, for instance, including household formation and dissolution as well as changes in occupation and income. Involuntary or forced moves are necessitated by events totally beyond the control of the household (Clark and Onaka, 1983). However, motives behind residential mobility are difficult to separate, and not easily to assign. In many cases several factors together lead to a move (Birg and Flöthmann, 1992). Furthermore, it is useful to make a distinction between reasons for moving and reasons for moving to a specific residence (Ravenstein, 1885; Ravenstein, 1889; Rossi, 1955).

Concerning the spatial context, distance and direction of moves are of central interest (Wagner, 1990). Residential mobility is frequently described by the distance covered, differentiating between international, national, regional or local migration (Franz, 1984). Most residential moves are characterised by short distances (Blijie, 2005; Franz, 1984). This means that the further a region is away from another region, the less likely are moves occurring between these regions, and vice versa (Franz, 1984). The direction of moves is strongly influenced by the available information about potential destination areas. In general, the regions persons have the most information about are preferred (Franz, 1984).

2.3 Ownership of mobility tools

Mobility tools include driving licences and available cars as well as different public transport season tickets, such as national and regional tickets for different time periods and half-fare discount tickets as well as others. Through the ownership of those mobility tools people commit themselves to particular travel behaviours, as they trade large one-time costs for low marginal costs at the time of usage. Especially, there is a commitment to the usage of the corresponding modes of transport. Simma and Axhausen (2003) found that the ownership of the different mobility tools influences the usage of the same mode positively and the usage of the other mode negatively, whereby the effects on a particular mode are greater than the effects on the other mode. This means that the relationship between the private and public transport mode is a substitutive one, as the commitment to a mode is connected to relatively high costs (Simma and Axhausen, 2003). In general, the commitment to car availability is higher than that to season ticket ownership. In this context, the ownership of cars and the related commitment are widely covered in the literature (Bhat and Sen, 2006; De Jong, 1996; De Jong, Fox, Daly, Pieters and Smit, 2004; Hensher, 1998), whereas the commitment to public transport is seldom considered in studies, as they mostly only emphasise its supply. Models taking into account both the ownership of cars and the ownership of different public transport season tickets are rarer (Axhausen, Simma and Golob, 2001; Beige, 2004; Scott and Axhausen, 2006; Simma and Axhausen, 2003). Scott and Axhausen (2006) showed that there exists a strong correlation between car and season ticket ownership. They demonstrated that modelling one without accounting for the other is likely to lead to biased results and that it is necessary to consider the trade-offs made between cars and season tickets.

Different variables influence the ownership of the various mobility tools (Beige, 2004; Simma and Axhausen, 2003). The relationship between age and ownership is non-linear. Men are more likely to own driving licences and cars, whereas women show a higher public transport season ticket ownership. Education and employment status as well as income have positive effects on the driving licence and car ownership. In this context, car ownership responds more strongly to rising than to falling income, i.e., the effect of rising income on car ownership is not totally reversed, as income falls. This means that the elasticity with respect to rising income is significantly greater than the elasticity with respect to falling income. This hysteresis effect is caused by habit or resistance to change or by the tendency to easier acquire than to abandon habits to consume. Increasing income has given individuals the possibility of owning a car and the convenience of its use. This is difficult to give up, as people become accustomed to it. The acquisition of a car is seen as a luxury, but once acquired the car becomes a necessity, so that disposing of a car is much more difficult (Dargay, 2001). This is an indication of the difficulty of reducing car dependence in favour of other transport modes. In addition, there is evidence that the income elasticity is not constant, but instead declines with increasing income (Dargay, 2001). A higher income also promotes the ownership of public transport season tickets (Beige, 2004). The location of the place of residence influences the ownership in such a way that people living in more urban areas tend to have less cars and more public transport season tickets at their disposal, as they have better access to public transport in comparison to rural areas (Beige, 2004; Karlaftis and Golias, 2002). The location of opportunities for activities plays also an essential role (Lanzendorf, 2006). The commuting distance to the place of occupation has a positive influence on car availability and a negative influence on season ticket ownership (Karlaftis and Golias, 2002). The access to the various transport systems as well as their quality and quantity is of high importance. This concerns on the one hand the options, comprising the whole range of means of transport, and on the other hand the conditions, referring to the accessibility of the options in terms of prices, costs and timing (Kaufmann, 2002). Concerning the ownership of cars, Karlaftis and Golias (2002) showed that parameters describing the transport networks, their efficiency and service primarily affect the number of cars owned by a household. The decision to acquire the first car is mainly related to socio-demographic and socio-economic factors, such as age, gender, income and household structure, whereas the purchase of the second and third car is largely based on characteristics of the different transport systems. With the size of the household, i.e., the number of adults and children, the level of car ownership increases (Bjørner and Leth-Petersen, 2004; Karlaftis and Golias, 2002).

3 Life course

The study of human lives represents an enduring interest of sociology as well as of the developmental and behavioural sciences, reflecting social changes and their consequences for individuals.

3.1 Concept and structure of the life course

The life course is formed by individuals' choices, actions and constraints (Elder, 2000). Decisions are made to shape one's life optimally within the external situation and the general conditions, following subjective criteria (Blossfeld and Huinink, 2001). Blossfeld and Huinink (2001) describe the life course as a self-referential process. This means that a person acts on the basis of past experiences and resources. Therefore, earlier decisions influence later decisions and also the possibilities for later decisions (Kelle and Kluge, 2001). Mayer (1990) speaks in this context of an endogenous causal connection, where subsequent choices, objectives and expectations are affected by experiences, conditions and constraints undergone previously. This means that people's behaviour is explained by its continuity over the life time and by specific events that involve main changes (Jones *et al.*, 1983; Lanzendorf, 2003). Thereby, the various dimensions of life are linked together and, hence, need to be considered together. The history in any life domain is involved in the behavioural processes of a person.

Age gives structure to the life course. There are sensitive phases, events that occur too early or too late, so life time is a constraint for actions (Elder, 2000; Lück, Limmer and Bonß, 2006). At the same time, the life course of individuals is embedded in as well as shaped by the historical contexts and places they experience over their life time (Elder, 2000). The impact of societal changes is particularly strong for young adults, as they are in a rather formative period, being about to make important decisions concerning their whole life. They are old enough to be influenced and change their behaviour, but not old enough to have become committed to a residence, an occupation or a way of life (Ryder, 1965; Tuma and Hannan, 1984). The longer a person persists in an established state, the less flexible he or she becomes. Consistency through time is achieved by developing a behaviour which reuses past decisions in identical or similar situations and, hence, routinisation predominates (Ryder, 1965; Schönfelder, 2006). In this context, the concept of cohorts plays an important role. It assumes that the influence of general changes varies for people experiencing these changes in different phases of their life. In corresponding analyses successive cohorts are compared with one another (Ryder, 1965). A cohort is defined as a group of individuals who experienced the same event within the same time interval (Aufhauser, 1995; Ryder, 1965). In general, this event is the birth of a person. Other possibilities include the move out of the parents' house, entering or leaving a certain occupation, marrying and divorcing, the birth of children, etc. Figure 2 illustrates the temporal dimensions concerning the life course showing the different influences related to the age or the life time of individuals as well as related to the historical context (Brandtstädter, 1990).

Figure 2   Temporal dimensions concerning the life course

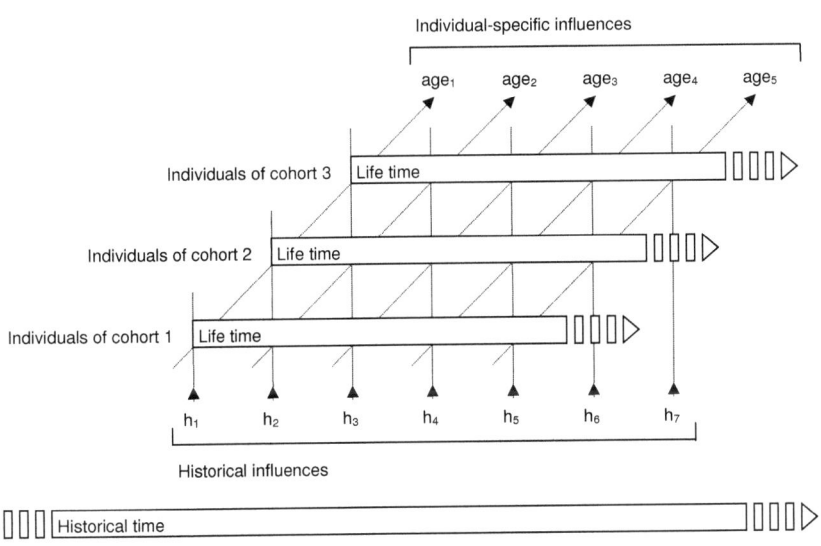

The structure of the life course is defined by the concepts of trajectories and transitions. The trajectories describe different domains of life, such as the social background, the family, education and employment, spatial mobility as well as health and diseases. The transitions represent changes in a state that are more or less abrupt, for example, marrying, divorcing, the birth of children, entering and leaving school, acquiring a job, retiring. Each transition is embedded in a trajectory that gives it specific form and meaning (Elder, 2000). As inter-mediary concept, the life course is seen as a sequence of events (Elder, 2000; Sackmann and Wingens, 2001). In this context, the timing of a life course indicates the points of time and the age at which transitions from one state to another state occur (Aufhauser, 1995). In addition, timing applies to the scheduling of multiple trajectories and their synchrony (Elder, 1995). The spacing describes the durations of certain states and the temporal gaps between events as well as rates of change. The sequencing shows the succession of different states (Elder, 2000). Sequencing is usually memory-endowed in the sense that a later state in the sequence remembers information of earlier states, which means that states are not independent (Sackmann and Wingens, 2001). So, it is worthwhile to understand an event as well as the history leading up to its occurrence (Box-Steffensmeier and Jones, 2004). Some decisions influence the whole life course, especially during sensitive phases, and have a high importance for the further development (Elder, 2000). Certain states tend to be more or less stable or more or less instable. In general, it is assumed that the longer a person remains in a certain state the less likely a change becomes, due to a stronger commitment to this situation (Elder, 2000).

The life course is described as a process with a continuous time axis and a multidimensional state space which is a set of mutually exclusive and collectively exhaustive states and represents the time varying characteristics of the individuals (Blossfeld and Huinink, 2001; Elder, 2000). Each domain of life forms a part of this process. They are linked together by time and age. During the life course the various dimensions have different priorities (Blossfeld and Huinink, 2001). Furthermore, the life course is embedded in the external conditions, i.e., the structuring influence of other people's life courses and society.

Lives are lived interdependently with so-called significant others, for instance, among members of a family and kin and friends, as experiences are shared through this network of social relationships (Elder, 1995; Elder, 2000; Keupp and Röhrle, 1987; Scott, 2000). Personal actions have consequences for others, and the actions of others impinge on the self (Elder, 1995). The connections between linked lives extend over the life span and generations, since these dynamics tend to persist from one generation to the next through a process of individual continuity as well as through intra- and inter-generational transmission (Elder, 2000).

3.2 Qualitative and quantitative approaches

Life course research primarily concentrates on the study of events in different domains, thereby embedding transitions into trajectories (Elder, 2000). In this context, connections and interrelations between various events are of interest as well as the delays occurring from one to the next change. Examining the life course as a whole rather than as single transitions leads to considerably higher complexity (Sackmann and Wingens, 2001).

In order to analyse people's life courses, there exist two approaches, a qualitative one and a quantitative one. Qualitative oriented techniques are used in biographical research (Elder, 2000). They are based on a self-reported narration of life, where individuals subjectively reconstruct their biography. In this context, not only occurring events and their timing are considered, but also the interpretation of the personal meaning as well as the self-reflexion of the respondents, in order to detect the reasons behind behavioural decisions (Blossfeld and Huinink; Elder, 2000). Corresponding data is collected retrospectively by means of qualitative interview methods. The narrative interview is most commonly applied, in which several stimuli are used with the aim of encouraging the narration by the interviewees about their life (Ohnmacht, 2006). This technique strongly relies on the individual's recall capacity. The relatively high expenses of collecting and analysing the data only enable the study and comparison of a few cases (Kelle and Kluge, 2001). This entails that generalisations are not possible. In contrast, the qualitative approach, therefore, represents a rather explorative method (Kelle and Kluge, 2001). The quantitative approach prevails in the so-called life course research. It conceptualises the life course of individuals as a sequence of events which denote transitions from one state to another, regarding the life course as objective history (Kelle and Kluge, 2001). In this context, life course research concentrates on measurable entities, applying statistical methods and estimating models, in order to determine the influence of time-independent and time-dependent covariates. An important focal point is the analyses of age, cohort and period effects by comparing different cohorts with one another.

It is possible to integrate both approaches, so that they complement one another. Kelle and Kluge (2001) as well as Sackmann and Wingens (2001) suggest that a small qualitative survey is used in an explorative way, while the results found this way are validated by means of a quantitative survey with a larger sample. Thereby, qualitative techniques are able to close gaps in the understanding of life courses by revealing the causal connection of inter-relationships observed in quantitative data.

3.3 Description of the life course

In the following, life courses of individuals are generally described, giving a short account of typical phases. Then, tendencies observed concerning long-term and mid-term mobility decisions during the life course are identified.

Heckhausen (1990) shows that, with respect to life time related norms, there commonly exists a high consensus among individuals about which behaviour and events are adequate and to be expected at which age (Aufhauser, 1995). However, increasingly individualisation processes are observed, where "normal" biographies become rather "elective", "reflexive", and "do-it-yourself" biographies (Beck and Beck-Gernsheim, 2002). Individualisation denotes the transition of the individual from heteronomy to autonomy, which manifests itself in a differentiation and pluralisation of life courses and styles (Beck, 1986). On the one hand, individualisation means the disintegration of the previously existing social order, for example, the increasing fragility of such categories as class and social status, gender roles, family, neighbourhood, etc. as well as the failure of tradition and inherited recipes for living to function. On the other hand, new demands, controls and constraints are being imposed on individuals, as modernity advances. Everything concerning life becomes increasingly decidable, and, at the same time, everything has to be decided by the individuals themselves. Thereby, the consequences are shifted from society onto the individuals. One of the decisive features of the individualisation process is that it not only permits, but also demands an active contribution by the individual, far more than earlier (Beck and Beck-Gernsheim, 2002). At the same time, the range of options widens, i.e., the choices among possibilities concerning the place of residence, education, employment, partnership, number of children, and so on increase (Beck, 1986). In this context, individuals have to be able to plan for the long-term and mid-term and to constantly adapt to changing conditions, thereby developing their own biography. Individualisation as well as related to it the opportunity and also the obligation to lead a life of one's own emerge, when a society becomes highly differentiated. It is a social condition which does not necessarily happen by choice (Beck and Beck-Gernsheim, 2002).

3.3.1 General description

The life course itself can be regarded as a contextual system (Mayer, 1990). A person's past affects his or her present, and his or her present affects his or her future (Ryder, 1965). Individuals seek coherence and continuity. So, an individual's life course and the successive events that constitute it are not random, but patterned.

The initial contribution to the design of a life time is made at conception, when the individual is provided not only with a fixed genetic constitution, but also, under ordinary circumstances, with parents to whom society assigns the responsibility for his or her early socialisation. Central characteristics present at birth are gender, family and kinship, birthplace and so forth. Perhaps the most important influence of status ascription on the future of an individual is the access to different amounts and kinds of formal and informal education, especially in German-speaking regions (Aufhauser, 1995; Ryder, 1965). Generally speaking, the life course is pre-structured by the milieu of origin in the first life decade and by selection processes concerning education and employment in the second life decade. Afterwards, situational decisions, such as the forming and dissolving of partnerships, the birth of children, moves as well as occupational transitions, primarily determine the directions into which different life courses develop. With increasing age the effect of the origin diminishes and life courses become more and more differentiated (Aufhauser, 1995). At the same time, the actions of an individual and those of others progressively reduce the degrees of freedom left to him or her throughout life. Facing various decisions among alternatives, an individual generally makes the choices somewhat congruent with the past, forming a growing commitment to a certain line of action (Ryder, 1965).

Concerning the familial development, the following phases are distinguished: marriage, pre-child, child-bearing, child-rearing, child-launching, post-child, and family dissolution (Wagner, 1990). Prolonging education generally leads to a delay in establishing a family of one's own. Furthermore, there is an increasingly longer period observable between forming a partnership and the birth of a child, in which women furthermore remain engaged in education or employment. A marriage is less frequently a reason for women to give up their occupation. Only the birth of a child is a decisive factor to make a break. The higher the occupational position reached, the longer women postpone childbearing and the sooner they start working again. Besides, birth related breaks are less frequently final and tend to be shorter. This is also strongly dependent on the available income of the household. Meanwhile, not the birth of the first child, but the birth of the second child becomes more important in shaping female life courses. For instance, suburbanisation is now more and more connected to the birth of the second child (Aufhauser, 1995). Divorces and remarriages recently increased, which means that two to three longer partnerships are common in the life of men and women. Early marriages as well as re-marriages are more prone to divorces. The presence of children is able to stabilise a marriage, but with diminishing effect over time (Aufhauser, 1995).

The time of entrance into the work force and the first work are important points in the life course, as they strongly shape later occupational careers. Overall, the share of labour participation continuously increases for both men and women in the middle life decades, whereas it declines on the verges of the employable age. Due to education, the entrance into employment occurs later, while retirement takes place earlier (Aufhauser, 1995). Occupational mobility is caused rather by the availability of vacancies than by changes in the qualification of the individuals. In the majority of occupations a steadily upward progression of status occurs throughout most of the life span, in form of a rise in income, prestige and power along an approximately age-graded continuum (Ryder, 1965). Prolonged membership in such structures decreases the probability of individual transformation. For instance, older employees become more and more adjusted to their work, turning towards tasks that are congenial to them.

3.3.2 Places of residence and moving behaviour

Persons and households need different housing at different points in life, for instance, using the age of the head of the household as a proxy for those different points (Morrow-Jones and Wenning, 2005).

There are three concepts implicitly or explicitly used in research on residential mobility over time. With increasing order of sophistication these are the housing career, the housing life cycle and the housing life course (Morrow-Jones and Wenning, 2005).

The housing career describes the sequence of dwellings that persons and households occupy over the course of their lives (Clark *et al.*, 2003). Usually, housing careers have a relatively simple structure and are characterised by stability over long stretches. In general, they show an upward trend in terms of size, number of rooms, quality, location, tenure, price and costs of housing. The change in price and costs paid for housing is used as a proxy for a number of other characteristics (Morrow-Jones and Wenning, 2005). Mobility is much higher among renters than among owners. Therefore, the change from renting to owning is a crucial transition in this sequence. In the early years of the housing career rental housing predominates (Clark *et al.*, 2003). The development of the housing career is primarily determined by the income level and income growth of persons and households (Clark *et al.*, 2003). For instance, low-income and moderate-income renters move to more comfortable quarters, better-off renters save to become first-time buyers, and first-time buyers trade-up to bigger and better homes. People move up this hierarchy, as they acquire the resources to do so. And the higher the income and the growth of this income, the earlier they reach a higher stage in the housing hierarchy. The top of the housing career is frequently reached between the ages of 40 and 50 years (Clark *et al.*, 2003). Thus, this approach mainly relies on the characteristics of the head of the household (age, marital status, occupation and income) to explain and predict housing choices. Housing careers illustrate the fact that household mobility decisions result in a series of incremental housing changes from a baseline stage toward something that approximates household satisfaction (Kendig, 1984).

Rossi (1955) showed that the major function of residential mobility lies in the process by which families adjust their housing to the housing needs, requirements or preferences that are generated by shifts in family size and composition that accompany life cycle changes. The housing life cycle is a model which ascribes changes in the housing requirements and preferences of households to different stages, progressing from the initial family formation (partnership, marriage), expansion (birth of children), contraction (maturation of children) and dissolution (break-up, divorce, death of a partner or spouse). Any transition to a new stage in the life cycle increases a household's potential for moving, because housing characteristics, such as size, number of rooms, tenure or location, no longer meet the needs, requirements and preferences of the family. When this mismatch occurs, families are increasingly likely to decide to move, in order to adjust their housing to achieve a better fit with household needs, requirements and preferences and thereby to improve their living conditions (Morrow-Jones and Wenning, 2005; Wagner, 1990). The housing life cycle model predicts that households move up in size and price of housing, as the adult members of the household age and become more established in their careers and their income goes up. In addition, as the family size increases and the children in the household age, they need more space. A move down in this view is also associated with ageing, as children leave the parents' house, parents need less space. Later on, people retire and the income decreases as well as infirmities rise, so they need a more accessible and manageable house requiring less maintenance. In this context, marital status, presence and age of children as well as the family structure are typically used as proxies for the family life cycle stages, explaining a household's mobility and housing decisions (Morrow-Jones and Wenning, 2005).

The housing life course is more flexible and now the dominant model in residential mobility research. This approach acknowledges the family as one of several factors affecting an individual's life (Morrow-Jones and Wenning, 2005). Whereas the life cycle model represents family formation and development as a fixed sequence of static stages, the life course model views marriage, divorce, parenthood, and so on, as events in an individual's family career or trajectory. Further trajectories include the housing career, mentioned above, the education and employment career, etc. These various trajectories are closely interrelated with one another and transitions often occur simultaneously (Lelièvre and Bonvalet, 1994). The housing career again describes a household's residential history which indicates the housing units occupied by a household over a specific period of time as well as the corresponding timing (Wagner, 1990). Concerning the residential durations, it is found that the longer a household lives in a unit, the less likely it is to move, because over time the household develops strong economic, social and emotional ties that increase the costs of moving (Morrow-Jones and Wenning, 2005). This means that the rate of change is negatively influenced by the duration (Wagner, 1990).

At the same time, residential mobility strongly depends on earlier behaviour (Wagner, 1990). The propensity to move seems to be "inheritable". High mobility of the parents during an individual's childhood leads later in life to higher than average mobility (Courgeau, 1985). An individual's "independent" residential trajectory starts in general with the move out of the parents' house, implying the formation of a new household (Lelièvre and Bonvalet, 1994). The transition of leaving the parents' house increasingly tends to take place earlier in life and is more frequently followed by shorter or longer periods of living alone or together with other people, before cohabiting with a partner (Aufhauser, 1995; Spiegel, 1992). Marriage leads to a period of relative stability in the residential trajectory. Married people are less mobile than unmarried people. In a French study, for example, the propensity to move is reduced by about one third after marriage (Courgeau, 1985). The birth of children and, therefore, an increase in the household size also tends to have a dampening effect on mobility (Lelièvre and Bonvalet, 1994), whereas the departure of children from the household induces an increase in residential mobility to more appropriate housing (Courgeau, 1985).

Besides the family career, the occupational career has a significant influence on the moving behaviour of individuals. Education leads to a higher level of spatial mobility (Weißhuhn and Büchel, 1992). To a lesser extent this also applies to employment, however, this effect is not very clear. More important is the influence of the occupational status on mobility (Courgeau, 1985). Migration is not in general linked to increases in income (Wagner, 1990). However, changes in the places of employment usually affect the income development positively. In this context, the income grows with the distance between the old and the new work place (Weißhuhn and Büchel, 1992).

The residential behaviour varies during the life course, but these patterns are relatively stable over time. Between the ages of 18 and 25 years people are the most mobile, moving several times within a short period of time (Birg and Flöthmann, 1992). This is also the age span, when they usually leave their parents' house (Wagner, 1990). Afterwards, spatial mobility decreases, soon reaching a stage of relative stability in the housing situation, but it slightly increases again at the beginning of the seventh life decade (Wagner, 1990). At the same time, the reasons for moving change. Education related reasons are only during a relatively short period of any importance. However, this is the time, when most moves during the life course take place. At the beginning of the occupational career, changes in employment are especially linked with spatial mobility, when it is most profitable (Wagner, 1990; Wagner, 1992). After a phase of occupational consolidation between the ages of 30 and 35 years, employment related reasons considerably decline (Birg and Flöthmann, 1992). The importance of moves connected to the family trajectory continuously rises with increasing age. Housing adjustments are a central motivation for residential mobility over all age groups, while neighbourhood adjustments become more important later in life (Clark and Onaka, 1983).

In addition, residential mobility is closely related to the situation on the housing market (Aufhauser, 1995; Clark *et al.*, 2003). Hence, the succession of dwellings occupied over life time inevitably varies between cohorts which face different housing market conditions (Kendig, 1984).

3.3.3 Ownership of mobility tools

Another life course trajectory refers to mobility and is directly linked to an individual's travel behaviour, including the availability of a driving licence and a car, the ownership of public transport season tickets as well as the actual travel patterns (Lanzendorf, 2003; Lanzendorf, 2006). Travel behaviour is characterised by habits and routines and tends to stay unchanged (Lanzendorf, 2006; Prillwitz *et al.*, 2006). Therefore, the mobility trajectory is described by periods of relative stability over time on the one hand and events that change behaviour or at least have a potential to change behaviour on the other hand. These events are so-called key or life events which are defined as major events in a personal and familial life, such as marriage or divorce, a move, a change in occupation and the corresponding location as well as changes in health (Van der Waerden, Timmermans and Borgers, 2003; Verhoeven *et al.*, 2005). The (anticipated) occurrence of a key event leads individuals to intensively reconsider their current travel behaviour and to consciously reflect their decisions, as habits and routines are broken or at least weakened. Some key events alleviate certain constraints (Verhoeven *et al.*, 2005). Other key events change the space-time context within which travel decisions are made. For example, moving implies that the spatial-temporal configuration of the current activity space shifts. Consequently, accessibility, travel distances and timing change. This also applies to alterations in the place of occupation. Key events influence the number of available alternatives and their characteristics as well as the attitude of individuals towards these alternatives. The existence and strength of this influence depends on personal and familial characteristics, e.g., age, gender, occupation, income and household structure, as well as on other characteristics, e.g., infrastructure, the transport systems, safety, weather conditions, etc. (Van der Waerden *et al.*, 2003). In the literature, a number of significant key events are identified. Among them, the most important ones seem to be residential relocations as well as changes in occupation and related income variations (Prillwitz *et al.*, 2006; Scheiner, 2006; Stanbridge and Lyons, 2006). Another important key event is the birth of children, as families adjust their activities and travel patterns (Prillwitz and Lanzendorf, 2006; Verhoeven *et al.*, 2005).

Overall, the ownership of mobility tools is relatively stable over longer periods of time (Axhausen and Beige, 2003; Simma and Axhausen, 2003). This is especially true for the ownership of cars. Bjørner and Leth-Petersen (2005) give two possible explanations for the strong persistence of car ownership over time. On the one hand, past car ownership has impacts on preferences or constraints that influence future car ownership. This is described as "true" state dependence. On the other hand, persons or households differ in certain unobserved characteristics that influence their probability of car ownership. When these unobservables are correlated over time, past car ownership is correlated with future car ownership. This correlation derived from unobserved heterogeneity is generally labelled as "spurious" state dependence. Bjørner and Leth-Petersen (2005) found that "true" state dependence is more important and contributes more than unobserved time invariant heterogeneity.

Concerning mobility tool ownership during the life course, studies in Germany and Switzerland show that around 90% of the persons aged between 18 and 25 years have a driving licence and of these driving licence owners 90% have a car available to them (Beige, 2004; Heine, Mautz and Rosenbaum, 2001). Young households tend to acquire their first car rapidly. The acquisition of the second car is more gradual. In general, household car ownership increases over the life course up until the head of the household reaches about the age of 50 years, which coincides with the motorisation of grown-up children in the household. Thereafter car ownership declines (Bussière, Armoogum, Gallez, Girard and Madre, 1994; Dargay, 2001; Dargay, Hivert and Legros, 2006). This pattern closely follows that of the household income. At the same time, individuals and households do not respond instantaneously to changes in income, instead adjustments occur slowly over time (Dargay, 2001). This emphasises the importance of dynamics in travel behaviour. Studies based on dynamic formulations indicate that the long-run income elasticity is generally two to three times the short-run income elasticity (Bjørner and Leth-Petersen, 2005; Dargay, 2001). Simma and Axhausen (2001) also argue that some relationships between relevant variables explaining travel behaviour are time lagged, such as the travel distances by car availability or public transport usage by season ticket ownership. Bjørner and Leth-Petersen (2004) showed that the effect on car ownership is asymmetric in the sense that downwards adjustments are less frequent than upwards adjustments after changes in specific socio-demographic and socio-economic conditions occur.

With respect to the occurrence of key events, residential relocations from urban areas to rural areas influence travel behaviour towards a higher car dependency. The birth of children in a family also increases car dependency and car use (Heine *et al.*, 2001; Lanzendorf, 2006; Prillwitz and Lanzendorf, 2006).

One general hypothesis about mobility trajectories suggests that over the life course younger adults tend to be more open to change than the elderly. This means that travel habits and routines are relatively established for most of the people above the age of 35 years (Prillwitz and Lanzendorf, 2006).

4 Research questions

The aim of the thesis at hand is to explore long-term and mid-term mobility decisions using a longitudinal approach. In this context, developments over time and over the life course are described. The main focus lies on the residential locations and the moving behaviour on the one hand and the ownership of various mobility tools, considering both cars and different public transport season tickets simultaneously, on the other hand. In particular, the inter-relationships between these two aspects are analysed. At the same time, the personal and familial situation is taken into account and how corresponding events affect long-term and mid-term mobility.

Further explanatory variables incorporated in the analyses include socio-demographic and socio-economic characteristics, such as age, gender, occupation, income and the household structure, as well as variables describing the place of residence, the residential municipality and region. Some of these variables are strongly correlated with one another, e.g., the age, the square and the natural logarithm of the age. These different variables are used to find the best possible form of interrelationship between the dependent and independent variables. Cost related information, e.g., concerning the costs for living and for mobility, are not considered, as those are not available, especially not spatially and temporally disaggregated for the observed time period from 1985 to 2004.

On the one hand, the long-term and mid-term mobility decisions people make are considered. These are modelled using the various factors, mentioned above, as explanatory variables. On the other hand, the temporal dimension, with respect to durations, changes and delays between changes, is explored. In this context, the propensity to stay in a certain state or to change it is of interest. The events in the several dimensions of life are taken into account and their influence is investigated by means of event history modelling.

In the course of the analyses, it is difficult to make clear statements about the causal connection between the various aspects of long-term and mid-term mobility behaviour as well as about the influence of other dimensions of life, such as personal and familial events. In spite of the chronological order of two events the impact can be in the opposite direction, as events are anticipated in advance.

Nevertheless, by means of an improved understanding of long-term and mid-term mobility, it is possible to find ways to influence the ownership of the various mobility tools and, hence, change the actual travel behaviour, as ownership can be used as a proxy for the usage of the various mobility tools. The overall objective is to encourage a shift from the motorised private transport towards more sustainable means of transport, like public transport, cycling and walking. Thereby, the negative impacts related to motorised private transport, especially on the environment, such as the consumption of resources and energy, air pollution, noise production, permanent congestion and accidents, can be reduced.

However, it is important to notice that the explanatory variables considered in the analyses of the long-term and mid-term mobility decisions do not include those relevant for policy and planning instruments, such as costs. In addition, the ones considered are relatively difficult to influence.

5 Methodology

In order to address the research questions formulated about the long-term and mid-term mobility decisions during the life course, the following methods are applied, on the one hand univariate and multivariate discrete choice modelling as well as on the other hand duration modelling, also called event history modelling.

5.1 Discrete choice modelling

Discrete choice models are used to analyse decisions, where persons choose from a finite set of mutually exclusive and collectively exhaustive alternatives. These models are based upon the assumption of utility maximisation, i.e., an individual chooses that one alternative with the highest utility (Ben-Akiva and Lerman, 1985; Maier and Weiss, 1990). The complexity of human behaviour and incomplete information suggest that uncertainty needs to be taken into account. The random utility models reflect this uncertainty, as the utility U of the alternative j for the individual n is given by

$$U_{jn} = V_{jn} + \varepsilon_{jn}, \tag{5-1}$$

where V_{jn} is the deterministic part of the utility and ε_{jn} the probabilistic part, capturing the uncertainty. The systematic or observable component comprises the characteristics of the person (e.g., age, gender, education, employment, income), the situation determining the options available to the individual as well as the generic and specific attributes of the alternative

$$V_{jn} = \beta_{0jn} + \sum_{k} \beta_{k''jn} x_{k''jn} + \sum_{k} \beta_{k'jn} x_{k'jn} + \sum_{k} \beta_{kjn} x_{kjn}, \tag{5-2}$$

where β_0 is a constant, while the β's denote the parameters that are estimated. The variables x describe the person n, the situation and the alternative j. According to the concept of utility maximisation, the alternative i with the highest utility is chosen when

$$U_{in} \geq U_{jn} \tag{5-3}$$

or

$$V_{in} - V_{jn} \geq \varepsilon_{jn} - \varepsilon_{in}. \qquad (5\text{-}4)$$

Hence, only the differences between utilities are relevant. This means that the concept of utility is relative and not absolute (Ben-Akiva and Bierlaire, 1999).

5.1.1 The logit model

For the logit models, it is assumed that the error component of the utility function is independently and identically extreme-value or Gumbel distributed. The Gumbel distribution possesses computational advantages, since it is possible to calculate probabilities without numerical integration or simulation methods.

The generalised extreme value (GEV) model is derived from the random utility model by McFadden (1978). In the GEV model the probability for choosing alternative i from the choice set C_n is in general defined as

$$P(i \mid C_n) = \frac{e^{V_{in}} \dfrac{\partial G}{\partial e^{V_{in}}}\left(e^{V_{1n}},...,e^{V_{in}},...,e^{V_{Jn}}\right)}{\mu G\left(e^{V_{1n}},...,e^{V_{in}},...,e^{V_{Jn}}\right)}, \qquad (5\text{-}5)$$

where G is the generating function with the following properties (Ben-Akiva and Bierlaire, 1999)

1. G is non-negative,
2. G is homogeneous of the degree $\mu > 0$,
3. $\lim_{x_i \to \infty} G(x_{1n},...,x_{in},...,x_{Jn}) = \infty$,
4. and the kth derivative of G with respect to k distinct x_i's is non-negative if k is odd and non-positive if k is even.

The simplest and most widely used model in the GEV model family is the multinomial logit model (MNL) where the generating function is

$$G(x) = \sum_{i=1}^{Jn} x_i^{\mu} \qquad (5\text{-}6)$$

and the probability for choosing alternative i from the choice set C_n is represented by

$$P(i\mid C_n) = \frac{e^{V_{in}}}{\sum_{j\in C_n} e^{V_{jn}}}. \tag{5-7}$$

An important property of the MNL model is its independence from irrelevant alternatives (IIA). This property states that the ratio of the probabilities of two alternatives only depends on their utilities. The premise is that the decision of choosing between two alternatives is unaffected by the presence or absence of any other alternative in the choice set. Therefore, it is possible to add and remove alternatives without influencing the structure of the model. On the other side, the IIA-property is unable to reflect existing relationships among different alternatives in situations where random utilities are significantly correlated and share a number of unobserved attributes. In this context, the assumption of independence from irrelevant alternatives is not valid. There exist various ways for relaxing the IIA-criterion.

The nested logit model (NL) is an extension of the MNL model designed to capture correlations among alternatives. It is based on the nesting of the alternatives of the choice set C_n into subsets C_{mn}, such that

$$C_n = \bigcup_{m=1}^{M} C_{mn}. \tag{5-8}$$

The generating function G in the nested logit model is defined as

$$G(x) = \sum_{m=1}^{M}\left(\sum_{i=1}^{J_n} x_i^{\mu_m}\right)^{\frac{\mu}{\mu_m}}, \tag{5-9}$$

with $\mu > 0$, $\mu_m > 0$ and $\mu \leq \mu_m$. The scale parameters μ and μ_m reflect the correlations among the alternatives within the nest C_{mn}. Thereby, only their ratio is meaningful (Ben-Akiva and Bierlaire, 1999). In the case that μ equals μ_m, the nested logit model corresponds to a multinomial logit model. The probability for individual n to choose alternative i within the nest C_{mn} is given by

$$P(i\mid C_n) = P(C_{mn}\mid C_n)P(i\mid C_{mn}), \tag{5-10}$$

with

$$P(C_{mn} \mid C_n) = \frac{e^{\mu V_{C_{mn}}}}{\sum_{m=1}^{M} e^{\mu V_{C_{mn}}}} \quad (5\text{-}11)$$

and

$$P(i \mid C_{mn}) = \frac{e^{\mu_m V_{in}}}{\sum_{j \in C_{mn}} e^{\mu_m V_{jn}}}. \quad (5\text{-}12)$$

In the NL model, the alternatives within each nest correlate with one another, whereas correlations across nests are not taken into account. Therefore, it is essential that alternatives are grouped into clearly separated nests. Otherwise the NL model does not reflect the correlation structure properly.

In the cross-nested logit model (CNL), it is possible to assign an alternative to several nests, so that an alternative belongs to more than one nest (Bierlaire, 2001). Thus, more complex situations and overlapping of alternatives are captured. The generating function for the CNL model is defined as follows

$$G(x) = \sum_{m=1}^{M} \left(\sum_{i=1}^{Jn} \alpha_{im} x_i^{\mu_m} \right)^{\frac{\mu}{\mu_m}}, \quad (5\text{-}13)$$

with $\mu > 0$, $\mu_m > 0$, $\mu \leq \mu_m$, $\alpha_{im} \geq 0$ and $\sum \alpha_{im} > 0$. The parameter α_{im} is usually interpreted as the degree at which alternative i belongs to nest m. Therefore, a common normalisation of the model imposes that $\sum \alpha_{im} = 1$. However, Ben-Akiva and Bierlaire (1999) emphasise that this condition is a convenient normalization condition, but is not necessary for the model to comply with random utility theory. The probability that individual n chooses alternative i from the choice set C_n is represented by the sum of the single probabilities over all nests

$$P(i \mid C_n) = \sum_{m=1}^{M} P(C_{mn} \mid C_n) P(i \mid C_{mn}). \quad (5\text{-}14)$$

In Figure 3 the structures of the multinomial, nested and cross-nested logit models are compared using a small network with four links as example (Bekhor, 1999). In the network, there exist three routes between the origin and the destination, where both route 2 and route 3 use link B together. In the MNL model, it is not possible to treat link B separately. In the NL model each route belongs to a separate nest. This means that for larger networks, where several routes use the same link, links appear multiple times in the structure of the model. Thus, the overlapping problem is not solved very efficiently. In the CNL model all links are listed in the upper level and they can belong to different nests at the same time. Therefore, the structure of the last model remains with two levels relatively simple.

Figure 3   Comparison of the structures in MNL, NL and CNL models

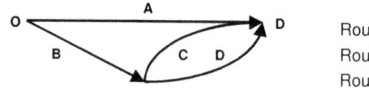

Route 1: Link A
Route 2: Links B and C
Route 3: Links B and D

Multinomial Logit Model   **Nested Logit Model**   **Cross-Nested Logit Model**

5.1.2 The probit model

The probit model is based on the assumption that the error component of the utility function is normally distributed (Ben-Akiva and Bierlaire, 1999). The main advantage of the probit model is its ability to explicitly capture all correlations among alternatives. It is completely general, because the variance-covariance matrix is arbitrary (Ortúzar and Willumsen, 2001). However, due to the high complexity of its formulation, the estimation and use of the probit model leads to problems (Cascetta, 2001; Vrtic, 2003).

5.1.3 Model estimation

The method most widely used for the estimation of the parameters in discrete choice models is the maximum likelihood method (Cascetta, 2001). Thereby, the values of the parameters are obtained by maximising the probability of reproducing the observed choices with the model.

The estimation of the various logit models is carried out using the software BIOGEME (Bierlaire Optimization Toolbox for GEV Model Estimation), version 1.4, which is developed for discrete choice models in general as well as for GEV models in particular (Bierlaire, 2005). The probit models are estimated with the software LIMDEP, NLOGIT 3.0.

5.2 Duration modelling

One approach for the analyses of life course dynamics includes the concepts of trajectory and transition (Elder, 2000). In this context, the life course is seen as a sequence of events. Thereby, it is of interest to understand an event and the history leading up to the occurrence of the event (Box-Steffensmeier and Jones, 2004). By means of event history modelling, differences in timing, rates of change and probabilities for the occurrence of certain events within a period of time are determined. At the same time, it is possible to study the influence of independent variables on the duration being the dependent variable.

An essential advantage of duration modelling over traditional linear regressions is its ability to account for the problem of right and left censoring. Censoring occurs, when information about durations is incomplete. This is the case, when preceding and subsequent events are unobserved, which means that the transition from one state to another is not made within the surveyed time. Problems arise, when uncensored and censored cases are treated equally, since the parameters in the duration model are maybe biased, and under- or overestimated. Furthermore, time-varying covariates, i.e., explanatory variables with values changing over time, are easily included in event history analysis (Box-Steffensmeier and Jones, 2004; Yamaguchi, 1991).

In the context of duration modelling, there exist different approaches. In parametric models the underlying hazard rate or transition rate, i.e., the rate at which events occur, is parameterised in terms of its probability distribution, e.g., Weibull, Gompertz, exponential, gamma, log-logistic and log-normal distributions (Allison, 1995). A semi-parametric alternative is represented by the Cox proportional hazard model (Cox, 1972; Cox, 1975). Thereby, it is not necessary to make assumptions about the particular distributional form of the duration times. This makes it preferable over its parametric alternatives (Box-Steffensmeier and Jones, 2004). In the Cox model the hazard rate for the ith individual is defined as follows

$$h_i(t) = h_0(t)e^{\beta' x_i}, \tag{5-15}$$

where $h_0(t)$ denotes the baseline hazard function and $\beta'x_i$ are the parameters and covariates. The hazard rate for the Cox model is proportional, as the hazard ratio for the two individuals i and j is written as

$$\frac{h_i(t)}{h_j(t)} = e^{\beta'(x_i - x_j)}, \tag{5-16}$$

which demonstrates that this ratio is constant over time (Box-Steffensmeier and Jones, 2004). The estimation method used in the Cox model is the maximum partial likelihood method and allows to estimate the parameters β' without specifying the baseline hazard function $h_0(t)$. This method is based on the assumption that the intervals between successive duration times contribute no information regarding the relationship between the hazard rate and the co-variates, but rather the ordered duration times (Box-Steffensmeier and Jones, 2004).

Event histories can consist of single events. On the other side, they can include multiple events of the same type or multiple events of different types. Cases where different kinds of events occur are often referred to as competing risks situations. There are many variants of competing risks models proposed in the literature (Allison, 1995; Box-Steffensmeier and Jones, 2004; Han and Hausman, 1990; Kalbfleisch and Prentice, 1980). A commonly applied approach is the latent duration time approach. It assumes that there are K $(k = 1,2,3,...,r)$ specific events and that there exists a potential or latent duration time associated with each event. The implementation of this model simply requires that K models with type specific hazards are estimated where all events other than k are treated as censored (Box-Steffensmeier and Jones, 2004). Thereby, the assumption is made that the K risks are conditionally independent. The latent variables approach is extended to both parametric and semi-parametric settings. Han and Hausman (1990) propose a flexible parametric proportional hazard duration model for competing risks, which permits unrestricted correlations among the risks. The specification of the model is flexible parametric in the sense that the baseline hazard is non-parametric, while the effect of the covariates takes a particular functional form, that is typically linear, but does not have to be. The underlying hazard model is based on a multivariate ordered probit model, which conceptually belongs to the group of discrete choice models. In this context, durations are treated as categorical by transforming the continuous intervals into an arbitrarily defined number of categories with a chosen class size, where each class needs to have at least two observations (Schönfelder, 2006).

For the estimation of the various duration models the software SPSS, version 14.0 is used.

6 Data

6.1 Data requirements

The analyses of long-term and mid-term mobility decisions require corresponding longitudinal data that describes people's life courses. Solely this kind of data enables the investigation of continuity and change over time (Ryder, 1965). In contrast to cross-sectional data which merely illustrates states and characteristics of an individual at one point in time, it is possible to explore not only statistical, but also causal relationships, as they are determined by a temporal sequence of cause and effect (Blossfeld and Huinink, 2001). Moreover, some relationships between relevant variables explaining behaviour are time lagged (Lanzendorf, 2003; Simma and Axhausen, 2001). These dynamic effects are only observable in longitudinal data and not in cross-sectional data (Dargay, 2001). Analyses of people's life courses do not consider a person's behaviour as isolated actions, but repeatedly in certain intervals (Ryder, 1965).

6.2 Data collection

Essentially, there are two ways of collecting longitudinal data. The most obvious and well-recognised method is to conduct a panel survey, in which the same sample of persons is asked about their respective current situation at several points in time to build up a series of observations. Data collected this way is very reliable, since events are observed, as they happen and, hence, inaccuracies due to memory loss are reduced (Diekmann, 1995; Lanzendorf, 2003; Zumkeller, Madre, Chlond and Armoogum, 2006). However, panel surveys are difficult and expensive to carry out as well as rather effort and time consuming, due to the long durations required for data collection (Scott and Alwin, 1998). Normally, it takes several years before it is possible to analyse long-term and mid-term effects (Lanzendorf, 2003). Therefore, panels do not represent a very flexible method. Another problem is imposed by panel attrition or mortality, which means that with each wave a number of participants leaves the panel and needs to be replaced by new participants (Lanzendorf, 2003). As a result, only a certain share of the initial respondents is still part of the panel in the long run. Thus, the reliability of the data is impaired. The second method approximating a panel survey is to use a retrospective approach that relies on individual's recall capacity and, therefore, is subject to the limitations of human memory. With increasing amounts of time elapsed since an event, the amount of information retained decreases in a logarithmic relationship (Brückner, 1994; Hollingworth and Miller, 1996; Lanzendorf, 2004). People tend to remember major events, such as residential moves or personal and familial events, better. Therefore, those are used as support for the memory by further linking different dimensions of life together and in doing so placing single events into a larger context (Brückner, 1990). Experiences from Hollingworth and Miller (1996) showed that a retrospective survey proved to be a favourable alternative to a panel survey. They tested it as a tool for collecting longitudinal data on residential mobility and found that people's ability to recall prior residential mobility decisions and housing details is generally good. Brückner (1994), Klein and Fischer-Kerli (2000), Lanzendorf (2004) and Peters (1988) also argue that a retrospective approach is feasible and suitable for important events of the life course that respondents are able to remember well. By adequate techniques, it is possible to improve the recall capability of the respondents, for instance, by starting with the most recent event and then tracing events backwards in time as well as by describing the event to be recollected relatively detailed (Ladkin, 2002). At the same time, retrospective data is easier, cheaper and faster to obtain than panel data (Gärling and Axhausen, 2003). Retrospective surveys allow for observing longer time spans than usually are feasible with panel surveys, whereas panel data is able to cover a broader range and more detailed information. More subjective concerns, such as attitudes, aspirations, expectations, motivations, opinions and the interpretation of events, are to a greater extent exposed to substantial bias by the respondents when recalled retrospectively than so-called hard facts (Scott and Alwin, 1998; Klein and Fischer-Kerli, 2000). There exist only a few studies that systematically compare the quality of retrospective data to prospective data, because its assessment requires corresponding panel data which is often missing (Dex, 1991; Klein and Fischer-Kerli, 2000; Peters, 1988; Scott and Alwin, 1998). Hence, little information about the reliability is available.

In order to collect the life course data concerning the long-term and mid-term mobility decisions, a retrospective survey covering the 20 year period from 1985 to 2004 is carried out. The survey is conducted using a written self-completion questionnaire which is sent out by mail. One reason for choosing this procedure is due to the relative complexity of the survey. In this way respondents have more time, quiet and privacy when answering the questionnaire, remembering and recollecting their past, possibly looking up documents if necessary. In addition, it is less demanding to obtain a larger sample, as expenses and costs tend to be in general lower in comparison to face-to-face and telephone interviews (Diekmann, 1995). At the same time, the influence of interviewers ceases to be an issue (Bird, Born and Erzberger, 2000). However, on the other side, it is not possible to offer immediate assistance to the respondents, in the case that problems with understanding and filling in the questionnaire occur. And corresponding queries on the part of the surveyed persons require some effort from their side. Generally, questions are more frequently not answered and the response overall is lower in mail surveys (Bird *et al.*, 2000; Diekmann, 1995). Dillman holds a contrary view, suggesting a tailored design method which takes into account different survey situations, thereby improving the general response (2000).

6.2.1 Questionnaire

Descriptions of survey instruments for the collection of life course data are difficult to find, especially when the survey is not conducted as face-to-face or telephone interview, but as mail survey instead (Bird *et al.*, 2000). In this case a survey instrument is needed that on the one side allows to record all essential information about people's life courses in high quality and as exactly as possible, but on the other side keeps the complexity within a certain limit and does not make excessive demands on the respondents, in order to also ensure a good response (Bird *et al.*, 2000).

The written self-completion questionnaire used here consists of two parts, a household form and a person form. The household form asks for the current address, a short description of all persons living in the household and the household income. In the person form socio-demographic and socio-economic characteristics of the respondents are collected. The essential part of this form is a multidimensional life course calendar for the years from 1985 to 2004. Such a calendar was used previously in life course research to retrospectively record family, education and employment histories as well as courses of disease (Bird *et al.*, 2000; Freedman, Thornton, Camburn, Alwin and Young-DeMarco, 1988). A life course calendar is based on a complete visual reconstruction of the past. So, a plain and compressed picture of the respondents' own life comprising several dimensions is developed which is also interesting and motivating for them to recover. Linking the various aspects together supports their recollection, as associations are formed (Brückner, 1990; Freedman *et al.*, 1988). At the same time, the graphic representation of the life course increases the quality and accuracy of the data, since inconsistencies in the timing of events between different dimensions become easier to detect. Furthermore, the life course calendar permits a comfortable handling of the complexity of the information and a rather straightforward recording of relatively detailed sequences of events in comparison to the conventional question-response format. Besides, it is a very flexible survey instrument (Bird *et al.*, 2000). The calendar itself is a matrix with a horizontal time axis for the observed time period from 1985 to 2004 with semi-annual precision. The six-month-intervals are chosen, because this time unit is small enough to ascertain the sequence and relation of events (Freedman *et al.*, 1988). But at the same time, it is necessary to consider the amount of detail as well as accuracy and time distinctiveness with which respondents are able to remember. Bird *et al.* (2000) made the experience that specifications on a semi-annual basis are feasible without larger difficulties. Along the other axis of the calendar the different items of the retrospective survey are arranged vertically. For the 20 year period, information about important events of the personal and familial history, such as the move out of the parents' house, marriages, divorces, births, deaths and retirements, is collected. Dex (1991) as well as Freedman *et al.* (1988) found that these events are more readily remembered and, therefore, are able to provide important reference points for the timing of other events. In addition, the household size and type are surveyed. Another dimension covers the moves and corresponding places of residence asking for the address, size and costs of the accommodation. Furthermore, the respondents are asked to indicate their changing ownership of cars and different public transport season tickets, such as national annual tickets, regional annual and monthly tickets and half-fare discount tickets. Data on the places of education and employment, on the main mode of transport for the commuting trip as well as on the personal income is also collected for the period from 1985 to 2004. Respondents are asked to enter in each case the duration by marking the beginning and the end. A line provides a record in a simple dichotomy and in parts a state (kind of event, places of residence, education and employment) needs to be specified. For better comprehension and understanding, an example of an already filled in life course calendar is included in the questionnaire.

Appendix A shows the questionnaire, including the household form and the person form. Each household receives two person forms that are to be filled in by persons aged 18 years and older. The time required to fill in the questionnaire depends very strongly on the frequency of changes occurring within the different dimensions of the life course during the observed 20 year period and amounts, based on own experiences, from 30 to 90 minutes.

6.2.2 Sampling

The survey is carried out in a stratified sample of municipalities in the Canton of Zürich, Switzerland, taking into account different spatial and transport system defined municipality types (Beige and Axhausen, 2005). In Figure 4 the location of the eleven study areas (Bassersdorf, Bülach, Dietlikon, Dübendorf, Horgen, Kloten, Rümlang, Wallisellen, Wangen-Brüttisellen, Winterthur and Zürich) is shown which focus on the Glattal region lying north of the City of Zürich. This focus is due to the initial cooperation with the project "Infrastructure, accessibility and spatial planning" of the Network City and Landscape. The objective of this project is to develop a dynamic urban simulation model for the Glattal region (Löchl, Bürgle and Axhausen, 2007; Waldner, Löchl and Bürgle, 2005).

Figure 4 Location of the study areas

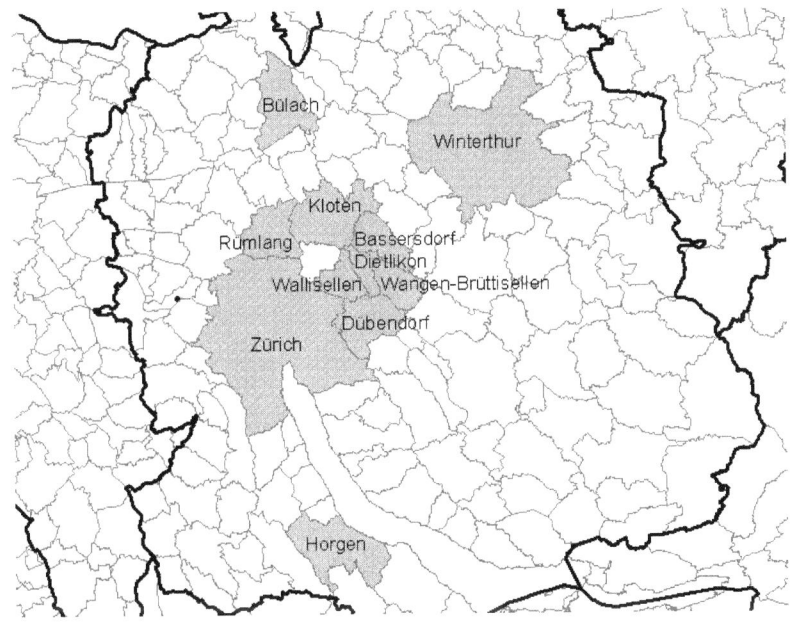

At the same time, predominantly households that moved within the last five years are sampled, including movers within the municipalities as well as arriving and departing residents. Therefore, about one fourth of the sampled households live in other Swiss municipalities. In the further analyses, this group is referred to as "Further addresses in Switzerland".

6.2.3 Description of the survey and response

In order to test the methodology, the feasibility of the survey and the questionnaire, a pre-test is carried out in the Zürich district 9 in October 2004. The main survey in the Zürich districts 3, 5, 11, 12 and the further ten study areas Bassersdorf, Bülach, Dietlikon, Dübendorf, Horgen, Kloten, Rümlang, Wallisellen, Wangen-Brüttisellen and Winterthur as well as in the other Swiss municipalities takes place from January to March 2005. The questionnaire, together with a self-addressed envelope is sent out by mail to 300 households in the pre-test and to 3300 households in the main survey. After two and four weeks, a reminder follows.

In the pre-test the response rate amounts to only 19.9%, which is primarily due to the relative length and complexity of the questionnaire (Axhausen, 2007). Therefore, in the main survey the households are, if possible, contacted by telephone within three days after they received the questionnaire to briefly explain the purpose of the survey and to motivate participation as well as to reduce concerns and to receive a feedback from the respondents. For 61.3% of the households, the telephone number is known and 50.8% are reached in this way. In this group the response rate reaches 30.9%, whereas in the other group with only 14.6% significantly fewer households participate in the retrospective survey. Figure 5 shows the response rate of the different municipalities depending on whether the households are contacted by telephone or not.

Figure 5   Response rate in regard to contact by telephone at municipal level

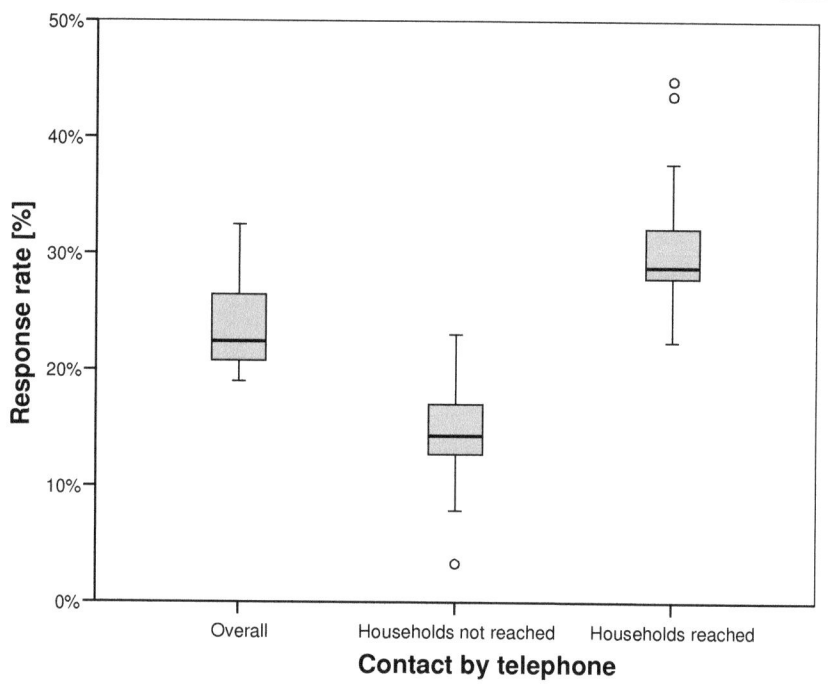

At the same time, significant differences between the two groups occur concerning the average duration of response, which is 16.8 days for the households reached by telephone and, thus, considerably lower than for the other group where it amounts to 21.2 days. Figure 6 shows the response rate over time in regard to the contact by telephone. Overall, the response is highest during the second week averaging 7.3%. Then, a decrease to 4.1% in the third week and an increase to 4.9% in the fourth week revealing a certain influence of the first reminder follow. Afterwards, the response declines continuously.

Figure 6   Response rate over time in regard to contact by telephone at municipal level

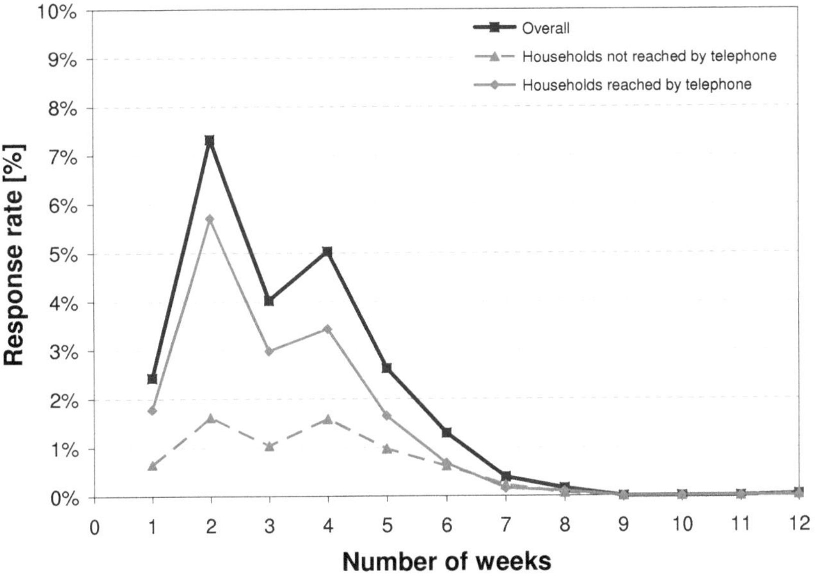

Considering both pre-test and main survey, the response rate is 23.1%. The data collected in the pre-test is also included in further statistical analyses, since the questionnaire is only slightly changed and extended for the main survey. Overall, 780 household forms and 1166 person forms are available. For one sixth of all the persons recorded in the household forms that are aged 18 years and older, the corresponding person forms are missing. In about two thirds of these 181 households more than two persons aged 18 years and older live, whereas for the other third only one of the two sent person forms is returned.

6.3   Representativeness of the data

In order to analyse the representativeness of the sample, the households and the persons that participate in the retrospective survey are compared to the entire population of Switzerland at the municipal level. In this context, data from the census of the year 2000, conducted by the

Federal Statistical Office, is used (2000). Furthermore, a comparison to the Swiss Household Panel data is carried out (2005).

6.3.1 Representativeness of the household sample

Table 1 and Table 2 show a comparison of the retrospective survey with the entire population by municipality with regards to size and type of the 780 households. Overall, the deviations between the two samples are relatively small. In the retrospective survey the households are on average slightly smaller, whereas the share of one-person households is lower with some more family and non-family households participating in the survey. Concerning the individual study areas, larger differences occur in Wangen-Brüttisellen and Zürich, where the households in the survey sample tend to be considerably larger. At the same time, there are fewer one-person households in comparison to the entire population. The opposite is evident in the municipalities Bassersdorf and Bülach.

Table 1   Household size by municipality

Municipality	Retrospective survey (2005)			Entire population (2000)		
	All persons	Adults	Children	All persons	Adults	Children
Bassersdorf	2.1	1.7	0.4	2.4	1.9	0.5
Bülach	2.1	1.5	0.6	2.3	1.8	0.5
Dietlikon	2.1	1.6	0.5	2.2	1.8	0.4
Dübendorf	2.0	1.7	0.3	2.1	1.7	0.4
Horgen	2.2	1.8	0.4	2.2	1.8	0.4
Kloten	2.1	1.8	0.3	2.1	1.7	0.4
Rümlang	2.3	1.8	0.5	2.1	1.7	0.4
Wallisellen	2.2	1.8	0.4	2.2	1.8	0.4
Wangen-Brüttisellen	2.9	2.1	0.8	2.4	1.8	0.6
Winterthur	2.4	1.9	0.5	2.2	1.8	0.4
Zürich	2.3	1.9	0.4	1.9	1.6	0.3
Further addresses in Switzerland	2.3	1.8	0.5	2.3	1.8	0.5
Overall	2.2	1.8	0.4	2.3	1.8	0.5

Table 2   Household type by municipality

Municipality	Retrospective survey (2005)			Entire population (2000)		
	One-person households	Family households	Non-family households	One-person households	Family households	Non-family households
Bassersdorf	37.2%	58.1%	4.7%	30.7%	66.0%	3.2%
Bülach	41.5%	52.8%	5.7%	33.0%	63.6%	3.4%
Dietlikon	35.9%	59.0%	5.1%	34.9%	62.2%	2.9%
Dübendorf	26.3%	68.4%	5.3%	39.8%	56.8%	3.4%
Horgen	32.6%	62.8%	4.6%	34.9%	62.0%	3.1%
Kloten	22.7%	72.7%	4.6%	40.1%	55.6%	4.3%
Rümlang	31.9%	66.0%	2.1%	37.0%	59.5%	3.6%
Wallisellen	28.2%	70.4%	1.4%	35.6%	59.7%	4.8%
Wangen-Brüttisellen	5.0%	95.0%	0.0%	28.9%	67.1%	4.0%
Winterthur	23.3%	73.3%	3.4%	38.9%	56.9%	4.2%
Zürich	31.1%	56.7%	12.2%	49.7%	43.0%	7.4%
Further addresses in Switzerland	28.7%	67.8%	3.5%	35.2%	60.7%	4.1%
Overall	30.4%	63.6%	6.0%	35.2%	60.7%	4.1%

In Table 3 and Table 4 all 1732 persons living in the households that are specified in the retrospective survey are described with regards to gender and age, respectively. For a small part of the household persons, primarily children, the corresponding data is missing. At the same time, the comparison with the entire population is shown. The shares of male and female persons in the survey sample are very similar to the census data of Switzerland. Concerning age, the persons living in the households are on average over three years younger in the retrospective survey than in the entire population, with a considerably higher share of persons aged from 20 to 64 years. This applies primarily to Rümlang, Dübendorf and the further addresses in Switzerland. Only in Kloten the household persons specified in the survey are older compared to the entire population of the municipality.

Table 3   Gender of all household persons by municipality

Municipality	Retrospective survey (2005)		Entire population (2000)	
	Male	Female	Male	Female
Bassersdorf	50.6%	46.1%	49.8%	50.2%
Bülach	45.4%	54.6%	48.9%	51.1%
Dietlikon	41.6%	54.5%	49.0%	51.0%
Dübendorf	51.4%	48.6%	49.5%	50.5%
Horgen	45.8%	54.2%	49.0%	51.0%
Kloten	48.8%	51.2%	51.0%	49.0%
Rümlang	52.8%	46.2%	49.9%	50.1%
Wallisellen	47.7%	50.3%	49.7%	50.3%
Wangen-Brüttisellen	51.7%	44.8%	50.8%	49.2%
Winterthur	48.6%	51.4%	48.4%	51.6%
Zürich	47.6%	50.2%	48.4%	51.6%
Further addresses in Switzerland	50.4%	48.6%	49.0%	51.0%
Overall	48.5%	49.9%	49.0%	51.0%

Table 4   Age of all household persons by municipality

Municipality	Retrospective survey (2005)			Entire population (2000)		
	Age in years	Aged below 20	Aged above 64	Age in years	Aged below 20	Aged above 64
Bassersdorf	36.9	20.2%	10.1%	37.5	23.4%	11.4%
Bülach	35.9	25.9%	11.1%	38.0	23.1%	12.2%
Dietlikon	35.0	20.8%	5.2%	39.7	20.5%	14.2%
Dübendorf	33.5	10.8%	10.8%	39.3	19.1%	13.3%
Horgen	37.1	18.8%	10.4%	40.0	20.7%	15.2%
Kloten	44.6	11.6%	14.0%	38.6	19.2%	12.5%
Rümlang	31.4	20.8%	3.8%	39.1	20.9%	14.4%
Wallisellen	43.9	17.2%	17.9%	40.9	19.1%	16.5%
Wangen-Brüttisellen	34.3	29.3%	3.4%	35.6	25.8%	8.0%
Winterthur	39.2	19.4%	16.7%	39.7	20.9%	16.5%
Zürich	35.2	19.0%	11.5%	41.3	15.8%	18.3%
Further addresses in Switzerland	33.3	21.6%	5.9%	39.2	22.9%	15.4%
Overall	35.9	20.1%	9.9%	39.2	22.9%	15.4%

Based on these results, the households in the retrospective survey are not weighted with respect to the Swiss population.

6.3.2 Representativeness of the person sample

Table 5 and Table 6 show gender and age of the 1166 persons participating in the survey concerning long-term and mid-term mobility. Furthermore, this sample is compared to the entire population in the different municipalities. In this context, only persons aged 18 years and older are taken into account. Regarding gender, the deviations to the entire population amount to only about 2%, though the differences are larger in some of the municipalities, especially in Dübendorf. Overall, the persons participating in the retrospective survey are slightly younger than the Swiss average. Noticeably the persons aged from 20 to 64 years are overrepresented in the sample. The largest differences concerning age occur again in Rümlang, Dübendorf and the further addresses in Switzerland.

Table 5   Gender of all persons by municipality

Municipality	Retrospective survey (2005)		Entire population (2000)	
	Male	Female	Male	Female
Bassersdorf	53.1%	46.9%	49.2%	50.8%
Bülach	43.5%	56.5%	48.5%	51.5%
Dietlikon	47.2%	52.8%	48.6%	51.4%
Dübendorf	59.3%	40.7%	49.0%	51.0%
Horgen	48.5%	51.5%	48.4%	51.6%
Kloten	48.6%	51.4%	50.9%	49.1%
Rümlang	55.2%	44.8%	49.5%	50.5%
Wallisellen	51.8%	48.2%	49.1%	50.9%
Wangen-Brüttisellen	54.3%	45.7%	50.7%	49.3%
Winterthur	50.0%	50.0%	47.8%	52.2%
Zürich	47.4%	52.6%	47.9%	52.1%
Further addresses in Switzerland	51.4%	48.6%	48.4%	51.6%
Overall	49.9%	50.1%	48.4%	51.6%

Table 6   Age of all persons by municipality

Municipality	Retrospective survey (2005)			Entire population (2000)		
	Age in years	Aged below 20	Aged above 64	Age in years	Aged below 20	Aged above 64
Bassersdorf	44.5	3.1%	14.1%	45.2	3.1%	14.4%
Bülach	45.8	0.0%	15.9%	45.6	3.0%	15.4%
Dietlikon	42.6	0.0%	7.5%	46.7	2.7%	17.4%
Dübendorf	38.0	0.0%	14.8%	45.7	2.3%	16.1%
Horgen	46.1	0.0%	14.7%	47.2	2.6%	18.7%
Kloten	47.5	0.0%	17.1%	44.9	2.3%	15.1%
Rümlang	39.1	0.0%	6.0%	46.2	2.6%	17.7%
Wallisellen	51.5	0.0%	24.1%	47.6	2.3%	20.0%
Wangen-Brüttisellen	46.3	0.0%	5.7%	43.9	3.1%	10.4%
Winterthur	46.7	0.0%	18.8%	46.9	2.7%	20.3%
Zürich	42.7	1.2%	15.9%	46.8	1.8%	21.3%
Further addresses in Switzerland	40.4	0.8%	7.8%	47.1	2.9%	19.3%
Overall	43.6	0.7%	13.6%	47.1	2.9%	19.3%

Furthermore, the residential mobility of the persons is analysed. Table 7 shows a description of the person sample with regards to gender, age and the place of residence five years ago as well as a comparison to the entire Swiss population of the year 2000. The largest deviations between the two samples occur in the group with the same address five years ago. Only about one fourth of the persons in the retrospective survey still live at the same place of residence, whereas this share amounts to over one half in the entire Swiss population. On the other side, persons living in another municipality are overrepresented in the survey sample. These differences are connected to the sampling of the households, which aimed for a higher share of households that moved recently. Overall, the deviations are relatively evenly distributed over both genders and the three age groups.

Table 7   Gender, age and place of residence of all persons five years ago

Gender	Age	Place of residence five years ago	Retrospective survey (2005)	Entire population (2000)
Male	From 18 to 39 years	Same address and same municipality	2.7%	7.3%
		Other address and same municipality	3.2%	3.6%
		Other municipality and same canton	7.2%	4.1%
		Other municipality and other canton	3.9%	2.0%
		Not specified	5.6%	2.1%
	From 40 to 59 years	Same address and same municipality	6.9%	12.2%
		Other address and same municipality	1.6%	2.4%
		Other municipality and same canton	5.1%	1.9%
		Other municipality and other canton	1.5%	0.7%
		Not specified	3.0%	0.9%
	60 years and more	Same address and same municipality	4.1%	8.7%
		Other address and same municipality	0.9%	1.1%
		Other municipality and same canton	1.8%	0.5%
		Other municipality and other canton	0.9%	0.2%
		Not specified	1.6%	0.5%
Female	From 18 to 39 years	Same address and same municipality	4.2%	6.8%
		Other address and same municipality	3.1%	3.5%
		Other municipality and same canton	7.6%	4.3%
		Other municipality and other canton	4.7%	2.1%
		Not specified	6.4%	2.2%
	From 40 to 59 years	Same address and same municipality	5.7%	12.5%
		Other address and same municipality	2.0%	2.2%
		Other municipality and same canton	4.3%	1.7%
		Other municipality and other canton	1.2%	0.6%
		Not specified	1.7%	0.8%
	60 years and more	Same address and same municipality	3.2%	11.4%
		Other address and same municipality	1.2%	1.7%
		Other municipality and same canton	1.7%	0.8%
		Other municipality and other canton	0.3%	0.3%
		Not specified	2.7%	0.9%
Overall		Same address and same municipality	26.8%	58.8%
		Other address and same municipality	11.9%	14.5%
		Other municipality and same canton	27.7%	13.3%
		Other municipality and other canton	12.6%	5.9%
		Not specified	21.0%	7.4%

Based on this comparison, a weighting of the retrospective survey sample is implemented at the person level using the three variables gender, age and the place of residence five years ago. The weights range from 0.2 to 3.6.

6.3.3 Comparison to the Swiss Household Panel

In the following, the sample of the retrospective survey is compared to the Swiss Household Panel (SHP). The principal aim of the Swiss Household Panel is to observe stability and change, in particular, the dynamics of changing living conditions in the population of Switzerland. It is a joint project of the Swiss National Science Foundation, the Swiss Federal Statistical Office and the University of Neuchâtel. The Swiss Household Panel exists since 1999. In the meantime the data of the first five waves from 1999 to 2003 are available. In the last wave 3289 of originally 5074 households participated. In total 8478 persons are surveyed, of which only 4900 persons are used in the comparison due to the age restriction of being 18 years and older as well as due to missing values. The two surveys are compared at the household and person level. A more detailed description of this comparison on the basis of a spatial classification into five different municipality types is found in Beige and Axhausen (2005).

Comparison of the household sample

Table 8 shows a description of different household characteristics for the retrospective survey one the one hand and the Swiss Household Panel on the other hand. The values in parentheses indicate the standard deviations of the mean values, which tend to be in general a little bit smaller in the survey concerning long-term and mid-term mobility. The households in the Swiss Household Panel are larger than the ones in the retrospective survey, with both more adults and more children living in these households. There are also considerably more families and fewer non-families. The households in the Panel have a slightly higher income at their disposal. At the same time, accommodations in the Swiss Household Panel have on average more rooms. Overall, fewer renters and more owners participate in the panel than in the retrospective survey. The costs for accommodation observed in the Swiss Household Panel are lower. In this context, it is necessary to take into account that in the retrospective survey the data describing the accommodation is missing for about one fifth of the households. A variance analysis shows that significant deviations between the two surveys occur with regards to the number of all persons, adults and children living in the households as well as with regards to the family and non-family households. Furthermore, the shares of renters and owners as well as the costs for accommodation differ considerably.

Table 8   Household and accommodation description

	Retrospective survey (2005)		Swiss Household Panel (1999-2003)	
Household size:				
All persons	2.2	(1.2)	2.5	(1.4)
Adults	1.8	(0.7)	1.9	(0.8)
Children	0.4	(0.8)	0.6	(1.0)
Household type:				
One-person households		30.4%		29.3%
Family households		63.6%		69.9%
Non-family households		6.0%		0.8%
Household income per month	8227 CHF	(4104 CHF)	8453 CHF	(6069 CHF)
Accommodation size:				
Rooms	4.0	(1.4)	4.2	(1.5)
Accommodation type:				
Rented		58.8%		54.2%
Owned		21.3%		44.7%
Accommodation costs:				
Rent per month	1582 CHF	(608 CHF)	1319 CHF	(564 CHF)
Rental value per year *	20647 CHF	(9197 CHF)	19121 CHF	(11251 CHF)

The values in parentheses indicate the standard deviations of the mean values.

* The rental value is defined as the equivalent rent, in the case that the accommodation would be rented instead of owned.

Comparison of the person sample

Table 9 describes the persons participating in the retrospective survey, using the weighted data, and the Swiss Household Panel. Concerning the shares of male and female persons as well as concerning the average age, the two samples are relatively similar. At the same time, more Swiss and less foreign nationals participate in the retrospective survey than in the Swiss Household Panel. With regard to occupation, the two samples differ only to a small extent. The share of employed persons is a little bit higher in the survey concerning long-term and mid-term mobility, whereas there are fewer persons in education. Overall, the monthly income is slightly higher. A variance analysis shows significant deviations between the two samples for nationality as well as for occupation (education, employment and home duties) of the persons.

Table 9   Person description

	Retrospective survey (2005) *		Swiss Household Panel (1999-2003)	
Gender:				
Male		48.3%		48.5%
Female		51.7%		51.5%
Age:				
Age in years	47.4	(16.6)	46.8	(16.6)
Aged below 20		0.7%		3.6%
Aged above 64		20.2%		16.8%
Nationality:				
Swiss national		84.2%		79.8%
Foreign national		15.4%		20.2%
Occupation:				
In education		3.0%		7.4%
Full-time employed		43.9%		42.5%
Part-time employed		25.4%		19.3%
Job-seeking		1.2%		1.6%
Home duties		6.4%		10.1%
Retired		20.0%		18.6%
Person income per month	5326 CHF	(3455 CHF)	5271 CHF	(4961 CHF)

The values in parentheses indicate the standard deviations of the mean values.

* The weighted data is used.

Table 10 describes the residential mobility of the two person samples. For both the retrospective survey and the Swiss Household Panel, the number of yearly moves as well as the time that has elapsed since the last move is shown. In this context, it is necessary to remember that in the retrospective survey the 20 years from 1985 to 2004 are considered, whereas waves of the Swiss Household Panel are conducted annually from 1999 to 2003. However, in both cases the last move before the observed time period is taken into account, where corresponding data is available. Overall, the comparison of the two samples shows that the persons in the Household Panel tend to move more often, whereas, at the same time, their last move dates further back. Differentiated by gender and age the same tendencies are noticeable, the only exception being persons aged from 40 to 59 years which have a higher moving rate in the retrospective survey than in the Household Panel. A variance analysis of the residential mobility indicates that the differences between the two surveys regarding the time since the last move are significant.

Table 10 Gender, age and residential mobility

Gender	Age	Retrospective survey (2005) *		Swiss Household Panel (1999-2003)	
		Number of moves per year	Time since the last move in years	Number of moves per year	Time since the last move in years
Male	From 18 to 39 years	0.181 (0.140)	5.1 (5.5)	0.203 (0.322)	10.3 (9.2)
	From 40 to 59 years	0.124 (0.102)	12.0 (12.1)	0.112 (0.126)	13.9 (10.0)
	60 years and more	0.045 (0.052)	18.8 (17.3)	0.061 (0.056)	24.5 (14.3)
Female	From 18 to 39 years	0.180 (0.129)	5.0 (5.7)	0.237 (0.383)	8.5 (7.6)
	From 40 to 59 years	0.116 (0.106)	12.5 (11.7)	0.100 (0.081)	14.8 (9.7)
	60 years and more	0.046 (0.047)	17.8 (14.5)	0.061 (0.066)	25.7 (15.1)
Overall		0.124 (0.119)	11.0 (12.4)	0.138 (0.239)	15.2 (12.5)

The values in parentheses indicate the standard deviations of the mean values.

* The weighted data is used.

Accordingly, Figure 7 shows the residential mobility for the male and female persons as well as for the three different age groups. Both genders do not vary significantly from one another, whereas the three age groups show clear differences. With increasing age persons tend to move less and to longer stay at a place of residence.

Figure 7 Gender, age and residential mobility

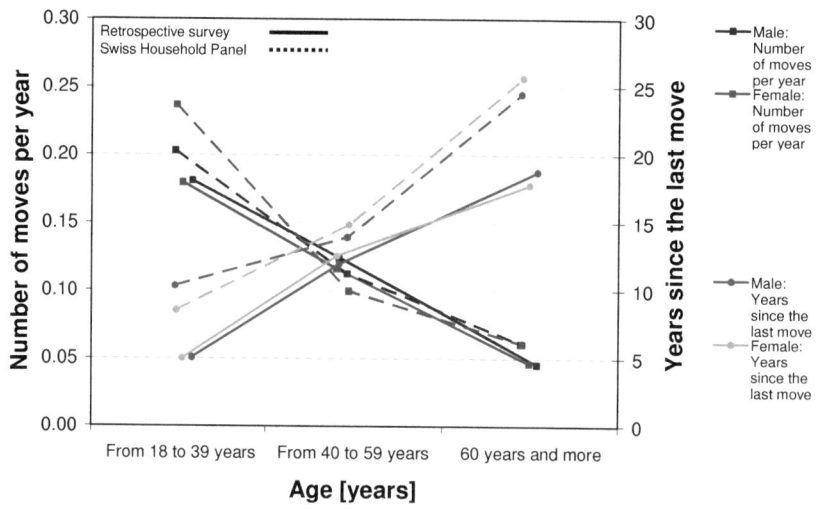

In Table 11 the results of a variance analysis of the number of annual moves as well as of the time since the last move are shown, using the weighted data. This analysis provides the possibility to study the effects of other variables on the means of these two variables. The variance is significantly influenced by the survey. Furthermore, gender and age play an important role. Regarding the time that has elapsed since the last move, the nationality as well as the occupation and the monthly income of the persons show high significance.

Table 11   Variance analysis of the residential mobility

Explanatory variable	Number of moves per year		Time since the last move in years	
	F-value	Significance	F-value	Significance
Survey: Retrospective survey	14.717	0.000	68.097	0.000
Gender: Male	18.711	0.000	8.027	0.005
Age in years	360.520	0.000	862.706	0.000
Nationality: Swiss national	0.219	0.640	16.159	0.000
Occupation:				
In education	0.095	0.758	0.188	0.665
Full-time employed	6.289	0.012	10.537	0.001
Part-time employed	2.743	0.098	10.724	0.001
Job-seeking	11.524	0.001	17.071	0.000
Home duties	2.556	0.110	8.902	0.003
Retired	8.737	0.003	6.135	0.013
Person income per month	0.807	0.369	40.069	0.000
Intercept	21.441	0.000	0.057	0.811
Number of observations		6030		5996
R^2 (adjusted)		0.107		0.304

Finally, Figure 8 shows the distributions of the two variables describing residential mobility for both surveys. Comparing the number of annual moves, the largest differences occur in the group of persons with less than 0.05 moves. There are considerably more of these persons included in the Swiss Household Panel, whereas the share of persons with 0.15 to 0.45 moves per year is lower. This is related to the fact that in the retrospective survey more mobile persons that moved within the last five years are sampled. Regarding the time that has elapsed since the last move, the share of persons that changed their place of residence within the last five years is nearly 20% higher in the retrospective survey. At the same time, more persons that moved the last time during the period from 15 to 25 years ago participate in the Swiss Household Panel. Probably a number of persons that moved during the period from 1999 to 2003 left the panel.

Figure 8   Residential mobility

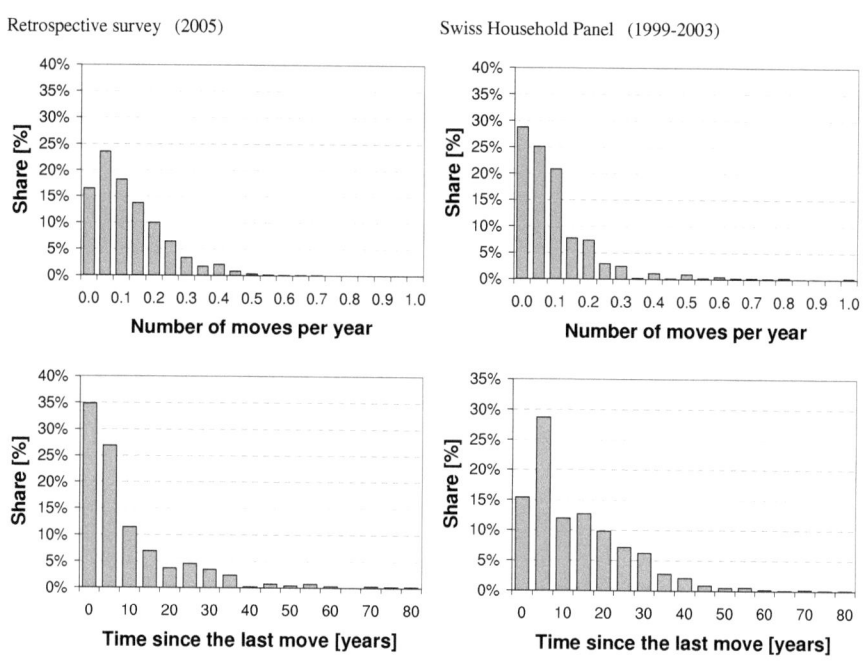

No further weighting of the retrospective survey sample based on the comparison with the Swiss Household Panel is implemented, since the two surveys are difficult to compare due to the different methodologies used. Furthermore, the sample size in the Swiss Household Panel is relatively small, increasing the chances of spurious differences.

7 Analyses for the year 2005

In the following chapter, the data is described and analysed for the year of its collection. Foremost, a spatial and transport system defined classification of municipalities which is used in the analyses is introduced. Subsequently, the households and persons are described in detail.

7.1 Spatial and transport system defined classification

The further statistical analyses are carried out using a spatial and transport system defined classification of the municipalities developed by the Federal Office for Spatial Development (2002). All Swiss municipalities are assigned to five different types. The first type covers the nine main centres (Basel, Bern, Geneva, Lausanne, Lucerne, Lugano, St. Gallen, Winterthur and Zürich). These are the core cities with more than 100000 inhabitants and more than 50000 workplaces. Types 2 and 3 comprise middle centres and ancillary centres of the main centres with and without access to the national railway network, respectively. In this context, access to the national railway network is defined as IC, IR, D and RX trains running at least every hour. The municipalities of the inner and outer agglomerations form the fourth type. The last type consists of the rural areas. In the following, this spatial and transport system defined classification is referred to as "ARE classification" using the German abbreviation of the Federal Office for Spatial Development. The classification is shown in Figure 9 for Switzerland and in Figure 10 for the study areas of the retrospective survey. As can be seen, none of the study areas is a rural area.

Table 12 shows the share of households and persons participating in the retrospective survey by the ARE type of the residential municipality. About 34% of the households and about 39% of the persons live in the main centres. Then, the middle and ancillary centres as well as the agglomeration municipalities follow. The rural areas are only represented with 10%. This last group is composed of households and persons that moved within the last five years from one of the study areas into another municipality. Therefore, this type shows a considerably lower share in comparison to the entire population of Switzerland, where the rural areas account for approximately 28% of all households and 31% of all persons. On the other side, the shares of households and persons living in the main centres are substantially higher in the retrospective survey compared to the Swiss average of 20% and 17%.

Figure 9　Spatial and transport system defined classification (ARE) of all Swiss municipalities

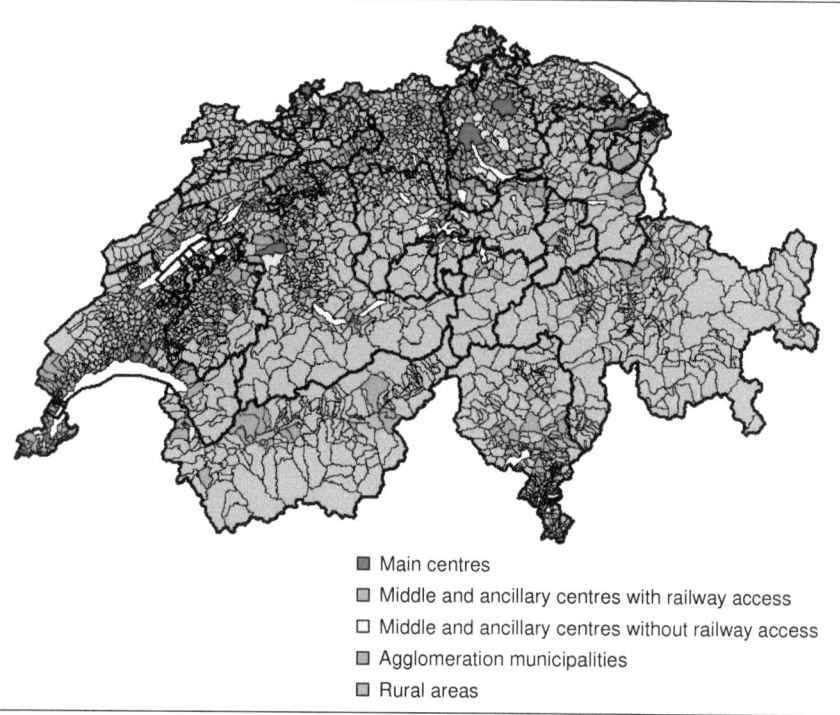

Figure 10   Spatial and transport system defined classification (ARE) of the study areas

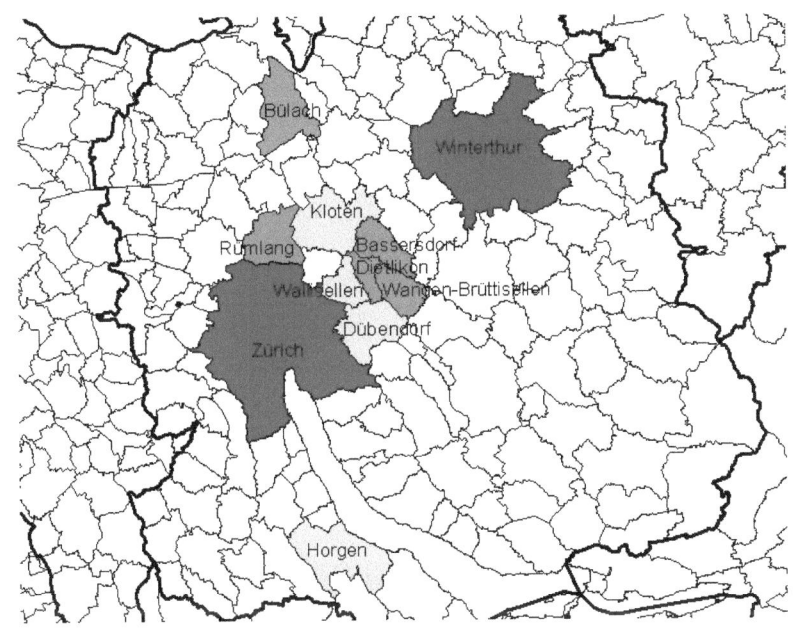

Table 12   Shares of households and persons by ARE classification of their place of residence (2005)

Spatial and transport system defined classification	Households	Persons
Main centres	33.5%	39.4%
Middle and ancillary centres with railway access	8.5%	7.8%
Middle and ancillary centres without railway access	21.3%	24.9%
Agglomeration municipalities	27.3%	18.3%
Rural areas	9.5%	9.6%
Overall	100.0%	100.0%

The following description of the data at the household and person level uses the spatial and transport system defined classification depicted above as its basis.

7.2 Description of the households

Table 13 and Table 14 show the size and type of the households with regards to the five different spatial and transport system defined municipality types. The values in parentheses indicate the standard deviations of the mean values. The mean household size is 2.2 persons, with 1.8 adults and 0.4 children in the households. The households in middle and ancillary centres are the smallest, whereas in the rural areas the number of persons is the highest. About 30% of the households participating in the retrospective survey are one-person households, about 64% family households and about 6% non-family households. Regarding the shares of families and non-families, significant differences occur between the municipality types. The share of non-family households is highest in the main centres. The most family households live in the rural areas, where, at the same time, the share of one-person households lies considerably below the sample average. The opposite tendency is visible in the middle and ancillary centres with railway access.

Table 13 Household size by ARE classification (2005)

Spatial and transport system defined classification	All persons		Adults		Children	
Main centres	2.3	(1.2)	1.9	(0.8)	0.4	(0.9)
Middle and ancillary centres with railway access	2.1	(1.2)	1.6	(0.7)	0.5	(0.8)
Middle and ancillary centres without railway access	2.1	(1.1)	1.7	(0.7)	0.4	(0.8)
Agglomeration municipalities	2.3	(1.2)	1.8	(0.7)	0.5	(0.8)
Rural areas	2.5	(1.2)	2.0	(0.7)	0.5	(0.9)
Overall	2.2	(1.2)	1.8	(0.7)	0.4	(0.8)

The values in parentheses indicate the standard deviations of the mean values.

Table 14 Household type by ARE classification (2005)

Spatial and transport system defined classification	One-person households	Family households	Non-family households
Main centres	31.0%	58.3%	10.7%
Middle and ancillary centres with railway access	39.4%	54.5%	6.1%
Middle and ancillary centres without railway access	28.9%	68.7%	2.4%
Agglomeration municipalities	30.5%	65.7%	3.8%
Rural areas	23.0%	72.9%	4.1%
Overall	30.4%	63.6%	6.0%

In Table 15 and Table 16 all household persons specified in the household form of the survey are described with regards to gender and age, respectively. For a small part of the household persons, the corresponding data is missing. Overall, slightly more women and girls than men and boys live in the households. This does not apply to the rural areas and to a lesser extent to the agglomeration municipalities. On average the household persons are nearly 36 years old, with 20% aged below 20 years and 10% aged above 64 years. In the rural areas and agglomeration municipalities they are youngest, with the lowest shares of persons aged 64 years and older.

Table 15   Gender of all household persons by ARE classification (2005)

Spatial and transport system defined classification	Male	Female
Main centres	47.9%	50.3%
Middle and ancillary centres with railway access	46.8%	53.2%
Middle and ancillary centres without railway access	48.0%	51.2%
Agglomeration municipalities	49.3%	48.6%
Rural areas	50.8%	47.5%
Overall	48.5%	49.9%

Table 16   Age of all household persons by ARE classification (2005)

Spatial and transport system defined classification	Age in years		Aged below 20	Aged above 64
Main centres	35.8	(20.9)	19.3%	12.7%
Middle and ancillary centres with railway access	34.6	(19.8)	24.5%	8.6%
Middle and ancillary centres without railway access	40.8	(20.2)	16.5%	13.9%
Agglomeration municipalities	33.9	(18.5)	20.7%	5.5%
Rural areas	32.9	(18.4)	24.6%	5.5%
Overall	35.9	(19.9)	20.1%	9.9%

The values in parentheses indicate the standard deviations of the mean values.

Table 17 shows the monthly gross income per month of the households. In this context, the data for 10.1% of the households is missing. Partly these missing items are corrected by the information specified in the life course calendar for the end of the year 2004, when the person forms of all adult members of the household are available. The still missing data for 5.8% of the households is imputed with SPSS, version 14.0, using the EM-method (expectation-maximisation method), which estimates the missing data in an iterative process on the basis of other variables. These variables include the number of all persons living in the household, the numbers of the male, grown-up and employed persons as well as the type of the residential municipality according to a classification into 13 different spatial types (Federal Office for Spatial Development, 2002). The mean value of the monthly household income is 8227 CHF with a standard deviation of 4104 CHF. A variance analysis shows that the income of the households is significantly different for the five municipality types. The mean value for the main centres lies considerably below the average of the survey sample. In this context, it is important to consider that in the main centre Zürich not all districts are taken into account, but only the districts 3, 5, 9, 11, 12 which are not necessarily in all aspects representative for Zürich.

Table 17   Household income per month by ARE classification (2005)

Spatial and transport system defined classification	Household income per month	
Main centres	7412 CHF	(3956 CHF)
Middle and ancillary centres with railway access	8170 CHF	(4389 CHF)
Middle and ancillary centres without railway access	8710 CHF	(4247 CHF)
Agglomeration municipalities	8716 CHF	(3964 CHF)
Rural areas	8657 CHF	(4082 CHF)
Overall	8227 CHF	(4104 CHF)

The values in parentheses indicate the standard deviations of the mean values.

Table 18 gives a short description of the accommodations the households reside in. It covers the number of rooms, the shares of rented and owned accommodations as well as the corresponding costs. For about one fifth of the households, the data concerning the accommodation is missing, and, therefore, the specified percentages do not sum up to 100%. On average the households have four rooms at their disposal. The accommodation size varies significantly for the five spatial and transport system defined municipality types. In the main centres the accommodations tend to be slightly smaller, whereas in the rural areas the mean value lies considerably above the average. In the household sample 59% rent and 21% own their accommodation, while for the remaining share of 20% the data is missing. The main centres are with shares of 70% for renters and 7% for owners very different from the other types. In contrast, approximately one third of the households in the rural areas and agglomeration municipalities own their accommodation. The mean costs amount to a rent of 1582 CHF per month and a rental value of 20647 CHF per year. It is necessary to take into account that the number of observations is not very large. This applies especially to owners in the sample.

Table 18 Accommodation size, type and costs by ARE classification (2005)

Spatial and transport system defined classification	Rooms	Rented	Owned	Rent per month	Rental value per year
Main centres	3.5 (1.3)	70.1%	6.5%	1513 CHF (590 CHF)	20533 CHF (9787 CHF)
Middle and ancillary centres with railway access	4.4 (1.6)	54.5%	25.8%	1573 CHF (554 CHF)	25772 CHF (18369 CHF)
Middle and ancillary centres without railway access	4.0 (1.2)	56.6%	24.1%	1655 CHF (642 CHF)	22297 CHF (8758 CHF)
Agglomeration municipalities	4.2 (1.4)	52.1%	31.0%	1646 CHF (619 CHF)	19411 CHF (6235 CHF)
Rural areas	4.8 (1.5)	47.3%	35.1%	1543 CHF (606 CHF)	18346 CHF (7897 CHF)
Overall	4.0 (1.4)	58.8%	21.3%	1582 CHF (608 CHF)	20647 CHF (9197 CHF)

The values in parentheses indicate the standard deviations of the mean values.

In Table 19 the shares of households owning one or more of various vehicles as well as the corresponding mean number and standard deviation are represented. These vehicles include cars, operable bicycles, motorcycles with more than $125cm^3$ and small motorcycles with less than $125cm^3$. Overall, about 75% of the households have a car at their disposal, 70% bicycles, 12% motorcycles and 6% small motorcycles. 9% of the households do not possess any of these vehicles. With regards to the ownership of cars, the five municipality types differ significantly from one another. In the main centres the share of households with one or more cars is considerably lower, as is the number of cars per household. By contrast, over 90% of all households in the rural areas have a car at their disposal. At the same time, the availability of bicycles there is clearly above the sample average. This also applies to the middle and ancillary centres with access to the national railway network. Motorcycles are most prominent in the rural areas as well as in the middle and ancillary centres.

Table 19   Household vehicles by ARE classification (2005)

Spatial and transport system defined classification	Cars		Bicycles		Motorcycles		Small motorcycles	
Main centres	0.7	52.9% (0.9)	1.5	65.9% (1.5)	0.1	7.7% (0.4)	0.1	5.0% (0.3)
Middle and ancillary centres with railway access	1.0	83.3% (0.6)	2.0	80.3% (1.5)	0.2	15.2% (0.4)	0.1	7.6% (0.3)
Middle and ancillary centres without railway access	1.2	83.7% (0.7)	1.5	71.1% (1.4)	0.2	11.4% (0.6)	0.1	7.2% (0.3)
Agglomeration municipalities	1.3	87.3% (0.8)	1.5	67.1% (1.5)	0.2	14.1% (0.4)	0.0	4.7% (0.2)
Rural areas	1.5	91.9% (0.8)	2.0	78.4% (1.6)	0.3	18.9% (0.6)	0.1	6.8% (0.3)
Overall	1.0	75.1% (0.9)	1.6	69.7% (1.5)	0.2	11.9% (0.5)	0.1	5.8% (0.3)

The values in parentheses indicate the standard deviations of the mean values.

7.3   Description of the persons

7.3.1   General description

Table 20 and Table 21 show the gender and age of the persons participating in the retrospective survey by ARE type of their residential municipality. The weighted sample is used. Overall, the shares of male and female persons are relatively balanced, also with regards to the different municipality types, the middle and ancillary centres with access to the national railway network forming the only exception. On average the respondents are 47 years old, with 1% being younger than 20 years and 20% being older than 64 years. Concerning age, the five municipality types differ significantly from one another. The mean value is lowest in the agglomeration municipalities, rural areas and middle and ancillary centres with railway access, due to a lower share of persons aged 65 years and older, whereas persons living in middle and ancillary centres without railway access lie considerably above the sample average.

Table 20   Gender of the persons by ARE classification (2005)

Spatial and transport system defined classification	Male	Female
Main centres	47.6%	52.4%
Middle and ancillary centres with railway access	44.9%	55.1%
Middle and ancillary centres without railway access	48.7%	51.3%
Agglomeration municipalities	49.4%	50.6%
Rural areas	50.6%	49.4%
Overall	48.3%	51.7%

Table 21   Age of the persons by ARE classification (2005)

Spatial and transport system defined classification	Age in years		Aged below 20	Aged above 64
Main centres	48.9	(18.4)	0.6%	26.4%
Middle and ancillary centres with railway access	44.7	(13.4)	0.0%	7.1%
Middle and ancillary centres without railway access	52.1	(15.9)	0.0%	26.8%
Agglomeration municipalities	41.9	(13.7)	1.2%	8.7%
Rural areas	42.4	(12.7)	2.5%	9.7%
Overall	47.4	(16.6)	0.7%	20.2%

The values in parentheses indicate the standard deviations of the mean values.

Table 22 shows the nationality of the persons participating in the retrospective survey by the spatial and transport system defined classification. Overall, there are 84% Swiss nationals and 15% foreign nationals in the sample. The five municipality types differ significantly. In the main centres the share of foreigners is slightly above the sample average, whereas in the middle and ancillary centres with railway access this share lies significantly below it.

Table 22   Nationality of the persons by ARE classification (2005)

Spatial and transport system defined classification	Swiss national	Foreign national
Main centres	83.3%	16.7%
Middle and ancillary centres with railway access	94.3%	5.1%
Middle and ancillary centres without railway access	80.9%	17.9%
Agglomeration municipalities	83.5%	16.3%
Rural areas	89.9%	10.1%
Overall	84.2%	15.4%

The values in parentheses indicate the standard deviations of the mean values.

Table 23 describes the occupation of the persons in the survey sample. Due to the possibility of choosing multiple answers, the shares sum up to over 100%. About 9% of the persons are in education, 44% full-time employed, 25% part-time employed and 2% seek a job. 14% are engaged in home duties and 24% are retirees. Education and job-seeking show the highest shares in the main centres, while employment lies below the sample average. In contrast, the share of employees is highest in the rural areas. This also applies to the house wives and house husbands in these areas, whereas the share of retirees is considerably lower than the average. The five municipality types thereby differ significantly with regards to the occupations "employment", "home duties" and "retirement".

Table 23   Occupation of the persons by ARE classification (2005)

Spatial and transport system defined classification	In education	Full-time employed	Part-time employed	Job-seeking	Home duties	Retired
Main centres	11.4%	37.2%	24.5%	2.4%	11.2%	30.8%
Middle and ancillary centres with railway access	8.6%	48.3%	27.6%	4.1%	15.9%	11.5%
Middle and ancillary centres without railway access	5.4%	42.3%	23.6%	1.8%	14.9%	30.7%
Agglomeration municipalities	8.7%	54.7%	27.7%	2.1%	13.9%	12.3%
Rural areas	12.1%	51.2%	27.1%	0.7%	23.5%	9.9%
Overall	9.3%	43.9%	25.4%	2.2%	14.1%	23.9%

Table 24 Education of the persons by ARE classification (2005)

Spatial and transport system defined classification	Hours per week	Travel distance in kilometres		Travel time in minutes	
		Private transport	Public transport	Private transport	Public transport
Main centres	20.3 (15.1)	6.8 (7.1)	7.6 (7.7)	17.2 (10.9)	30.7 (17.2)
Middle and ancillary centres with railway access	15.9 (19.3)	28.3 (52.9)	26.2 (45.8)	24.3 (30.2)	39.7 (45.1)
Middle and ancillary centres without railway access	23.2 (17.3)	11.7 (16.3)	12.3 (16.4)	19.9 (15.3)	34.5 (28.7)
Agglomeration municipalities	18.7 (16.0)	13.1 (14.5)	15.7 (15.9)	20.9 (15.8)	44.6 (30.0)
Rural areas	21.9 (17.7)	64.1 (74.8)	61.1 (64.8)	53.2 (50.6)	98.8 (84.1)
Overall	20.4 (16.0)	17.3 (35.7)	17.7 (32.1)	23.2 (24.9)	42.8 (42.8)

The values in parentheses indicate the standard deviations of the mean values.

Table 25 Employment of the persons by ARE classification (2005)

Spatial and transport system defined classification	Hours per week	Travel distance in kilometres		Travel time in minutes	
		Private transport	Public transport	Private transport	Public transport
Main centres	36.0 (13.3)	15.8 (76.2)	15.8 (72.4)	20.9 (41.3)	35.6 (57.5)
Middle and ancillary centres with railway access	35.8 (12.9)	13.1 (13.2)	15.2 (14.1)	18.4 (14.7)	43.0 (28.2)
Middle and ancillary centres without railway access	35.2 (14.5)	8.8 (9.5)	9.9 (9.7)	15.9 (11.5)	36.6 (22.3)
Agglomeration municipalities	35.4 (14.5)	10.1 (8.5)	12.1 (9.5)	17.5 (11.7)	40.7 (22.3)
Rural areas	36.2 (15.5)	21.1 (43.8)	23.6 (41.5)	21.3 (31.5)	54.0 (58.6)
Overall	35.7 (14.1)	13.2 (48.0)	14.4 (45.8)	18.8 (28.1)	39.6 (43.0)

The values in parentheses indicate the standard deviations of the mean values.

Table 24 and Table 25 show again for the persons in education and employment, respectively, the number of hours per week spent in these activities. Furthermore, the travel distances and travel times are presented for both private and public transport (Vrtic, Fröhlich, Schüssler, Axhausen, Dasen, Erne, Singer, Lohse and Schiller, 2005; Vrtic, Fröhlich, Schüssler, Axhausen, Schulze, Kern, Perret, Pfisterer, Schultze, Zimmerman and Heidl, 2005). The respondents in education are engaged in it for about 20 hours per week. The average air-line distance to the place of education amounts to approximately 13.3 kilometres. The travel distances are a little bit longer with 17 kilometres for private transport and 18 kilometres for public transport. The corresponding travel times show larger differences between the two modes with 23 minutes and 43 minutes, respectively. The five municipality types differ significantly regarding the travel distances as well as regarding the travel times. The values are considerably higher for the rural areas, whereas the trip to the place of education is shortest in the main centres. The employed persons in the sample work for nearly 36 hours per week. On average the place of employment is 10.0 kilometres away, with travel distances amounting to 13 kilometres for private transport and 14 kilometres for public transport. The trip takes 19 minutes and 40 minutes depending on the mode of transport. Again, the rural areas show the longest trips. But significant differences between the various municipality types occur only for the travel times by public transport.

Table 26   Mode of transport to education and employment by ARE classification only considering persons in education or employment (2005)

Spatial and transport system defined classification	Trip to the place of education				Trip to the place of employment			
	Private transport	Public transport	Bicycle	On foot	Private transport	Public transport	Bicycle	On foot
Main centres	16.0%	61.0%	13.1%	2.1%	26.5%	40.4%	18.5%	9.7%
Middle and ancillary centres with railway access	32.8%	39.5%	0.0%	27.8%	62.3%	17.7%	8.3%	11.8%
Middle and ancillary centres without railway access	36.0%	46.5%	0.0%	0.0%	55.0%	25.0%	5.8%	7.7%
Agglomeration municipalities	32.2%	47.6%	0.0%	2.8%	57.1%	27.8%	3.1%	6.3%
Rural areas	16.0%	80.7%	0.0%	0.0%	70.3%	14.4%	1.2%	9.2%
Overall	22.9%	57.5%	6.3%	3.5%	47.8%	29.2%	9.4%	8.6%

Table 26 illustrates the most frequently used mode of transport for the trip to the place of education as well as to the place of employment. In this context, 108 persons in education and 808 persons in employment are included. For some of these respondents, the data concerning the mode of transport is missing. Therefore, the specified percentages do not sum up to 100%. Overall, 23% of the respondents use private transport, 58% use public transport, 6% cycle and 4% walk to their place of education, whereas these shares amount to 48%, 29%, 9% and 9%, respectively, for the employed persons in the sample. In comparison to the other four municipality types, the main centres show significantly lower values for private transport and significantly higher values for public transport as well as a more frequent use of bicycles. In the rural areas, the share of public transport used for the trip to the place of education is comparatively high. This is probably connected to the service of school buses in these areas.

Table 27 shows the mean person income per month and the corresponding standard deviations. This data is derived from information about the gross income specified in the life course calendar. In this context, for 15.4% of the persons the data is missing, which is imputed using again the expectation-maximisation method in SPSS, version 14.0. The variables used in the iterative estimation process include gender, age, nationality, occupation, mobility tool ownership as well as the spatial type of the residential municipality (Federal Office for Spatial Development, 2002). Overall, the mean person income is 5534 CHF with a standard deviation of 3547 CHF. The values for the five spatial and transport system defined municipality types differ significantly from one another. The main centres are the only type with a value below the sample average, whereas the middle and ancillary centres show in comparison the highest incomes.

Table 27   Person income per month by ARE classification (2005)

Spatial and transport system defined classification	Person income per month	
Main centres	4997 CHF	(3130 CHF)
Middle and ancillary centres with railway access	5731 CHF	(3962 CHF)
Middle and ancillary centres without railway access	5692 CHF	(3609 CHF)
Agglomeration municipalities	5596 CHF	(3570 CHF)
Rural areas	4884 CHF	(3532 CHF)
Overall	5326 CHF	(3455 CHF)

The values in parentheses indicate the standard deviations of the mean values.

Table 28 and Figure 11 indicate how satisfied the respondents are with different items, such as health, accommodation, work, leisure time, the condition of the environment in the region as well as the state of life in general. In this context, the satisfaction is measured in a scale ranging from 1 for very satisfied to 10 for very unsatisfied. The table shows the mean values and the corresponding standard deviations. Overall, the satisfaction of the respondents ranges from very satisfied to at the least satisfied with relatively small standard deviations. The lowest satisfaction is indicated with regard to the environment. This especially applies to the first four spatial and transport system defined municipality types, whereas respondents living in rural areas are considerably more satisfied with their surroundings. Concerning the other items, the differences are not significant. In the rural areas persons tend to indicate a higher than average satisfaction with their health. In the main centres the evaluation of work and leisure time is better. On the other side, the middle and ancillary centres as well as the agglomeration municipalities show values below the average for the satisfaction with life in general.

Table 28   Satisfaction of the persons by ARE classification (2005)

Spatial and transport system defined classification	Health	Accommo- dation	Work	Leisure time	Environ- ment	Life in general
Main centres	3.0 (2.3)	2.7 (1.9)	2.8 (1.9)	2.5 (1.9)	4.1 (2.0)	2.6 (1.8)
Middle and ancillary centres with railway access	2.7 (1.8)	2.5 (2.0)	3.1 (2.1)	3.0 (1.8)	3.9 (1.9)	2.9 (2.0)
Middle and ancillary centres without railway access	3.0 (2.1)	2.8 (1.9)	2.9 (2.0)	2.9 (2.1)	4.1 (2.2)	2.8 (1.8)
Agglomeration municipalities	2.8 (2.1)	2.4 (2.1)	3.1 (2.1)	3.0 (2.1)	4.2 (2.2)	2.8 (2.0)
Rural areas	2.6 (2.0)	2.4 (2.0)	2.8 (1.9)	2.7 (2.0)	3.3 (2.0)	2.4 (1.8)
Overall	2.9 (2.1)	2.6 (2.0)	2.9 (2.0)	2.8 (2.0)	4.0 (2.1)	2.7 (1.8)

The values in parentheses indicate the standard deviations of the mean values.

Figure 11   Satisfaction of the persons by ARE classification (2005)

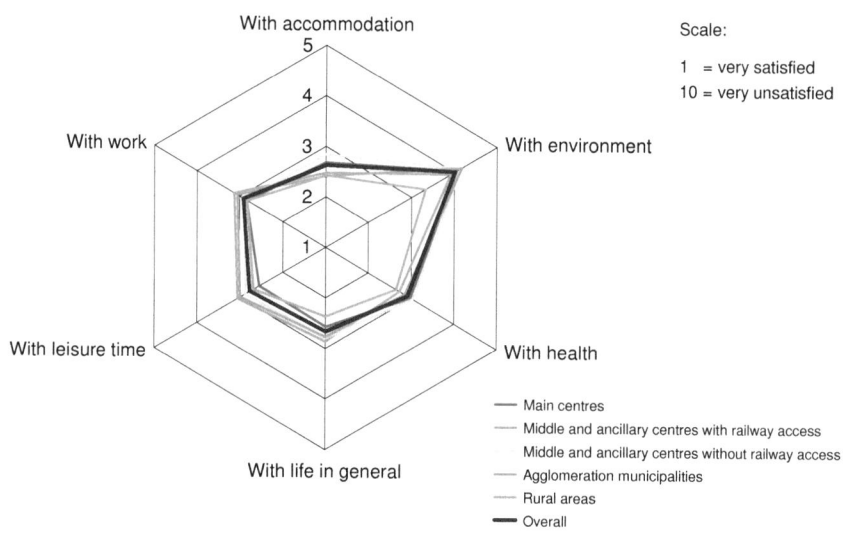

7.3.2 Places of residence and moving behaviour

Table 29 details the residential mobility as the number of annual moves and time that has elapsed since the last move by the ARE classification of the current place of residence, while Table 30 illustrates the reasons for the last move, where multiple answers are possible. Thereby, it is necessary to take into account that motives behind residential mobility are difficult to separate, and not easily to assign. In many cases several factors together lead to a move (Birg and Flöthmann, 1992). On average the persons in the survey sample moved about every 8 years with the last move occurring 11 years ago. The five municipality types differ significantly from one another. Persons living in the agglomeration municipalities tend to move most frequently. Personal and familial reasons are the first motivation concerning the last move with around 44%, closely followed by accommodation related reasons. About one fifth of the respondents specify the surrounding as cause for moving. Motives connected to education and employment show a share of about 20%. The vicinity to family and friends plays with 11% only a minor role. Significant differences between the diverse municipality types occur with regards to the personal and familial reasons, reasons related to the accommodation as well as with regards to the vicinity to family and friends.

Table 29   Residential mobility by ARE classification (2005)

Spatial and transport system defined classification	Number of moves per year	Time since the last move in years
Main centres	0.114 (0.124)	13.8 (15.7)
Middle and ancillary centres with railway access	0.138 (0.104)	8.6 (7.0)
Middle and ancillary centres without railway access	0.106 (0.102)	11.9 (10.8)
Agglomeration municipalities	0.165 (0.138)	5.5 (6.9)
Rural areas	0.121 (0.096)	9.7 (7.6)
Overall	0.124 (0.119)	11.0 (12.4)

The values in parentheses indicate the standard deviations of the mean values.

Table 30   Reasons for the last move by ARE classification (2005)

Spatial and transport system defined classification	Personal and familial reasons	Education and employment related reasons	Accommodation related reasons	Surrounding related reasons	Vicinity to family and friends	Other reasons
Main centres	41.0%	19.4%	38.1%	23.7%	6.1%	2.0%
Middle and ancillary centres with railway access	52.1%	18.4%	31.5%	20.2%	7.0%	8.9%
Middle and ancillary centres without railway access	39.2%	24.5%	51.7%	15.7%	14.6%	1.6%
Agglomeration municipalities	46.3%	15.8%	48.5%	19.9%	17.5%	0.7%
Rural areas	56.1%	15.0%	28.0%	20.8%	14.7%	0.2%
Overall	43.9%	19.5%	41.8%	20.5%	11.2%	2.0%

Table 31 shows the assessment of the respondents of the probability of changing their place of residence within the next year. Overall, more than half of the persons consider this as very unlikely and more than a quarter as unlikely, whereas only 11% and 7% indicate that a move is likely and very likely, respectively. Especially, in the middle and ancillary centres with access to the national railway network the occurrence of a move within the next year is rather unlikely in comparison to the other four municipality types.

Table 31   Probability for a move within the next year by ARE classification (2005)

Spatial and transport system defined classification	Very likely	Likely	Unlikely	Very unlikely
Main centres	6.0%	13.2%	25.8%	52.3%
Middle and ancillary centres with railway access	4.7%	2.6%	38.6%	53.4%
Middle and ancillary centres without railway access	5.9%	11.2%	34.4%	47.3%
Agglomeration municipalities	10.2%	9.1%	24.1%	54.9%
Rural areas	4.7%	14.7%	23.7%	55.4%
Overall	6.5%	11.3%	28.4%	51.9%

7.3.3 Ownership of mobility tools

Table 32 and Table 33 show the driving licence ownership, the car availability and the ownership of different public transport season tickets, respectively. 83% of the respondents own a driving licence for cars, which they acquired on average by the age of 21.8 years. At the same time, 57% have a car always, 18% partially and 22% never at their disposal. 3% of the persons participate in car sharing. The ownership of driving licences as well as the availability of cars shows significant differences with regards to the ARE types of the residential municipality. The main centres feature the lowest shares of driving licences and always available cars, whereas car sharing is most prevalent there. Concerning the public transport season ticket ownership, 12% of the respondents have a national annual ticket, 19% a regional annual or monthly ticket and 46% a half-fare discount ticket. With the exception of the last group, mobility tool ownership is significantly different between the five municipality types. The share of national and regional season tickets is considerably higher in the main centres. In the rural areas the ownership of half-fare discount tickets lies noticeably below the average of the survey sample.

Table 32 Driving licence ownership and car availability by ARE classification (2005)

Spatial and transport system defined classification	Driving licence ownership	Car availability			
		Always	Partially	Never	Car sharing
Main centres	70.6%	34.6%	24.1%	36.1%	5.1%
Middle and ancillary centres with railway access	90.6%	74.9%	12.8%	11.3%	1.0%
Middle and ancillary centres without railway access	87.6%	69.8%	12.7%	14.6%	2.9%
Agglomeration municipalities	91.4%	72.7%	11.6%	13.7%	2.0%
Rural areas	94.9%	72.2%	20.5%	5.5%	1.7%
Overall	82.6%	57.3%	17.7%	21.7%	3.3%

Table 33 Public transport season ticket ownership by ARE classification (2005)

Spatial and transport system defined classification	National ticket ownership	Regional ticket ownership	Half-fare discount ticket ownership
Main centres	18.8%	27.6%	48.6%
Middle and ancillary centres with railway access	11.3%	10.4%	47.0%
Middle and ancillary centres without railway access	9.2%	11.9%	43.7%
Agglomeration municipalities	4.4%	18.6%	46.8%
Rural areas	7.2%	8.1%	38.2%
Overall	12.0%	18.8%	45.9%

Overall, it is noticeable that the ownership of a driving licence is closely connected to the car availability. Over 90% of the driving licence owners have a car always or partially at their disposal. At the same time, a driving licence and an always available car have a negative influence on the ownership of national and regional season tickets, whereas the participation in car sharing involves a higher ownership of all the public transport season tickets considered. Between the national tickets on the one side and the regional and half-fare discount tickets on the other side, there exists a relationship of substitution, while the tickets within the last group tend to complement one another. These interrelationships between the different mobility tools are likewise visible in Table 34 and the corresponding Figure 12, where the public transport season ticket ownership is shown against car availability. Over 50% of the respondents with an always available car also own a public transport season ticket at the same time, whereas this share is nearly 80% for the respondents with a partially available car or without a car. Thereby, the ownership of national and regional tickets increases, while the share of persons only having a half-fare discount ticket at their disposal considerably decreases with a lower level of car availability. Nearly all persons participating in car sharing have a public transport season ticket. In this group the share of these tickets is the highest. Thus, car sharing is primarily used as a complement to public transport.

Table 34   Car availability and public transport season ticket ownership (2005)

Car availability	National ticket	No national ticket			
		Regional ticket		No regional ticket	
		Half-fare discount ticket	No half-fare discount ticket	Half-fare discount ticket	No half-fare discount ticket
Always	5.6%	5.8%	4.7%	36.8%	47.1%
Partially	14.0%	20.3%	10.1%	32.7%	22.9%
Never	24.7%	19.0%	9.0%	26.2%	21.1%
Car Sharing	28.0%	39.9%	0.0%	30.2%	1.9%
Overall	12.0%	12.3%	6.4%	33.6%	35.7%

Figure 12   Car availability and public transport season ticket ownership (2005)

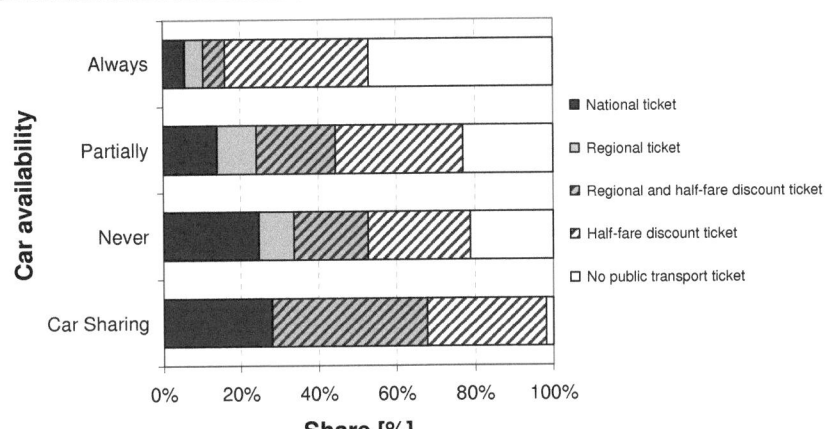

Discrete choice modelling for the ownership of mobility tools

Furthermore, univariate and multivariate discrete choice models are estimated, in order to analyse the decisions concerning the ownership of mobility tools during the life course. These models include binomial and multinomial logit models, nested and cross-nested logit models as well as a probit model.

For the estimation of the various discrete choice models, only 1154 of the 1166 respondents in the sample are included due to missing values. Table 35 gives a description of all the explanatory variables used in the models, showing their mean values, standard deviations, minima and maxima as well as the source of the data.

Table 35   Description of the explanatory variables (2005)

Explanatory variable	Mean value	Standard deviation	Minimum	Maximum	Source
Age in years	47.177	16.337	18.000	88.000	Retrospective survey (2005)
Gender: Male	0.486	0.500	0.000	1.000	Retrospective survey (2005)
Nationality: Swiss national	0.842	0.365	0.000	1.000	Retrospective survey (2005)
College or university degree	0.310	0.463	0.000	1.000	Retrospective survey (2005)
In education	0.094	0.292	0.000	1.000	Retrospective survey (2005)
In employment	0.700	0.459	0.000	1.000	Retrospective survey (2005)
Monthly income in 1000 CHF	5.340	3.468	0.055	16.000	Retrospective survey (2005)
Fuel price in CHF per litre (lead free 95)	1.487	0.015	1.390	1.553	Carle (2005)
Driving licence ownership	0.826	0.379	0.000	1.000	
Car availability: Always	0.573	0.495	0.000	1.000	
Car availability: Partially	0.177	0.382	0.000	1.000	Retrospective survey (2005)
National ticket ownership	0.120	0.325	0.000	1.000	
Regional ticket ownership	0.188	0.391	0.000	1.000	
Half-fare discount ticket ownership	0.459	0.499	0.000	1.000	
Number of persons in the household	2.556	1.123	1.000	9.000	Retrospective survey (2005)
Accommodation costs in 1000 CHF	1.616	0.605	0.400	6.944	Retrospective survey (2005)
Spatial and transport system defined classification:					
Main centres (referential category)	0.389	0.488	0.000	1.000	
Middle and ancillary centres with railway access	0.078	0.269	0.000	1.000	Federal Office for Spatial Development (2002)
Middle and ancillary centres without railway access	0.251	0.434	0.000	1.000	
Agglomeration municipalities	0.184	0.388	0.000	1.000	
Rural areas	0.096	0.295	0.000	1.000	
Population in the residential municipality in 1000 inhabitants	130.584	162.930	0.099	363.273	Federal Statistical Office (2000)
Population density in the residential municipality in inhabitants per square kilometre	20.879	15.448	0.084	111.997	

Table 36 presents the results of binomial logit models for the availability of cars and the ownership of the various public transport season tickets, only including variables with a level of significance higher than 0.10. Constants are not significant. As measure for the goodness of fit, the adjusted ρ^2 is shown in the table. It is calculated as follows

$$\rho^2 = 1 - \frac{L(\max) - K}{L(0)}, \qquad (7\text{-}1)$$

where $L(0)$ and $L(max)$ represent the initial and the final log-likelihoods, respectively, and K denotes the number of estimated parameters (Bierlaire, 2005). A rather low goodness of fit is observed only in the model for the half-fare discount ticket ownership, whereas in the other models ρ^2 is relatively high.

The age of the respondents has a positive effect on the existence of always available cars, for men to a greater extent than for woman. At the same time, men tend to own cars more frequently. This also applies to Swiss nationals in comparison to foreign nationals. Employment and a higher income increase the probability of an always available car as well. The fuel price shows a negative influence as well as the ownership of the different public transport season tickets. In the main centres car availability is considerably lower than in the other municipality types. Persons that only partially have a car available to them tend to be younger, more likely in education and less likely in employment. They live in bigger households, in on average less expensive accommodations and in more densely populated areas. The ownership of national tickets increases with increasing age, whereas regional ticket ownership is negatively influenced by age. Persons in education and employment own these two ticket types more frequently. The concurrent availability of a car leads to a lower ownership of all public transport season tickets. Appendix B 1 (Figure B 1-1, B 1-2, B 1-3) includes the corresponding graphs showing the influence of age and gender, the monthly income and the accommodation costs for the different utilities.

Table 36 Binomial logit models for car availability and public transport season ticket ownership (2005)

Explanatory variable	Car availability: Always	Car availability: Partially	National ticket ownership	Regional ticket ownership	Half-fare discount ticket ownership
Age in years	+ 0.080	+ 0.394	− 0.127		+ 0.058
Age in years squared	− 0.001	− 0.002	+ 0.002		− 0.001
Age in years natural logarithm		− 9.302		− 1.038	
Gender: Male	− 1.410				− 1.600
Age in years * Gender: Male	+ 0.045			− 0.008	+ 0.022
Nationality: Swiss national	+ 0.601		+ 1.281	− 0.484	+ 1.201
College or university degree					+ 0.541
In education		+ 0.658	+ 1.338	+ 0.810	
In employment	+ 0.959	− 0.671	+ 0.676	+ 0.614	
Monthly income in 1000 CHF				+ 0.069	
Monthly income in 1000 CHF squared					
Monthly income natural logarithm	+ 0.415				+ 0.006
Fuel price in CHF per litre (lead free 95)	− 5.080				
Driving licence ownership			+ 0.765	− 0.607	
Car availability: Always			− 2.047	− 0.864	− 0.342
Car availability: Partially			− 1.215		
National ticket ownership	− 2.234				
Regional ticket ownership	− 1.357				+ 1.100
Half-fare discount ticket ownership	− 0.856	+ 0.546		+ 1.125	
Number of persons in the household		+ 0.246			
Accommodation costs in 1000 CHF		− 2.844	− 1.062	+ 2.072	
Accommodation costs in 1000 CHF squared	+ 0.178		+ 0.260	− 0.501	
Accommodation costs natural logarithm		+ 3.342			− 0.382
Spatial and transport system defined classification:					
Main centres (referential category)					
Middle and ancillary centres with railway access	+ 2.246				
Middle and ancillary centres without railway access	+ 1.949				
Agglomeration municipalities	+ 2.167				
Rural areas	+ 2.031				
Population in the residential municipality in 1000 inhabitants	+ 0.002		+ 0.005	+ 0.003	− 0.003
Population density in the residential municipality in inhabitants per square kilometre		+ 0.021	− 0.045		+ 0.028
Number of observations	1154	1154	1154	1154	1154
ρ^2 (adjusted)	0.287	0.417	0.553	0.429	0.091

For all variables: Level of significance ≤ 0.10

In the following, only persons that are in education or employment are considered, which concerns 843 of the 1154 respondents. In Table 37 binomial logit models for the ownership of the various mobility tools are shown. Always available cars are more likely at the disposal of older, male, Swiss and employed persons with a higher income and higher accommodation costs. At the same time, the ownership of public transport season tickets has a negative influence. Outside of the main centres cars that are always available are more prevalent. The population of the residential municipality has a positive influence, while it is negative for the population density. Partially car availability decreases until the age of about 55 years with a negative mean elasticity of 1.718. Regarding gender, the mean elasticity is also negative with a value of 0.196. Being in education as well as the distances between the places of residence, education and employment show a positive effect. The ownership of half-fare discount tickets increases the probability for the ownership of partially available cars. The size of the household affects partial car availability positively, whereas the accommodation costs entail a negative mean elasticity of 0.129. Respondents living in municipalities with a higher population density tend to have more partially available cars at their disposal. With regard to the ownership of the various public transport season tickets, no clear tendencies are observable. The distance to the place of education increases the propensity of holding a national ticket, while the distance to the place of employment has a negative influence on the regional ticket ownership. The probability of owning a half-fare discount ticket rises with increasing age and is less common among men than among women. Driving licence ownership and car availability affect the holding of all the season tickets in a negative way. The different utilities due to age and gender, the monthly income and the accommodation costs are again presented in Appendix B 1 (Figure B 1-4, B 1-5, B 1-6). In the course of the estimation of further discrete choice models the partial sample of persons in education or employment is not considered separately, since the shown models are relatively similar to the entire sample and the number of observations is only about a fourth lower.

Table 37 Binomial logit models for car availability and public transport season ticket ownership only considering persons in education or employment (2005)

Explanatory variable	Car availability: Always	Car availability: Partially	National ticket ownership	Regional ticket ownership	Half-fare discount ticket ownership
Age in years		− 0.178			
Age in years squared		+ 0.002			
Age in years natural logarithm				− 1.549	+ 0.665
Gender: Male	− 1.864				− 0.659
Age in years * Gender: Male	+ 0.056	− 0.010			
Nationality: Swiss national	+ 0.934		+ 1.028		+ 1.167
College or university degree					+ 0.578
In education		+ 0.585	+ 0.889		
Distance between the place of residence and the place of education in kilometres		+ 0.015	+ 0.018		
In employment	+ 1.891				
Distance between the place of residence and the place of employment in kilometres		+ 0.006		− 0.003	
Monthly income in 1000 CHF	+ 0.092				+ 0.188
Monthly income natural logarithm					− 0.534
Driving licence ownership				− 0.643	
Car availability: Always			− 1.442	− 0.891	− 0.436
Car availability: Partially			− 0.710		
National ticket ownership	− 2.169				
Regional ticket ownership	− 1.399				+ 0.971
Half-fare discount ticket ownership	− 0.957	+ 0.385		+ 1.031	
Number of persons in the household		+ 0.236			
Accommodation costs in 1000 CHF	+ 0.794	+ 2.508	− 2.532		
Accommodation costs in 1000 CHF squared		− 0.920	+ 0.592		
Accommodation costs natural logarithm	− 0.599			+ 0.619	
Spatial and transport system defined classification: Main centres (referential category)					
Middle and ancillary centres with railway access	+ 1.953				
Middle and ancillary centres without railway access	+ 2.251				
Agglomeration municipalities	+ 2.390				
Rural areas	+ 2.055				
Population in the residential municipality in 1000 inhabitants	+ 0.007			+ 0.002	
Population density in the residential municipality in inhabitants per square kilometre	− 0.045	+ 0.022			

Table 37 is continued ...

Table 37 continued ...

Explanatory variable	Car availability: Always	Car availability: Partially	National ticket ownership	Regional ticket ownership	Half-fare discount ticket ownership
Number of observations	843	843	843	843	843
ρ^2 (adjusted)	0.296	0.398	0.553	0.354	0.100

For all variables: Level of significance ≤ 0.10

Discrete choice modelling for the ownership of mobility tools in groups

For further analyses, various mobility tool ownership groups are formed, covering all possible combinations, in order to consider the different mobility tools simultaneously. Table 38 shows the ownership of always and partially available cars and the different public transport season tickets, including national annual tickets (Nat T), regional annual and monthly tickets (Reg T) as well as half-fare discount tickets (HF T). The biggest group with 31% consists of persons only owning a car. Then, the group of car and half-fare discount ticket owners as well the group of car and national or regional ticket owners follow. Fewer than 5% of the respondents do not have any mobility tool at their disposal. The differences between the five municipality types are partly significant. In the main centres the share of persons without a car and, at the same time, owning public transport season tickets is considerably higher, whereas there are fewer persons with a car and no season tickets or half-fare discount tickets. The opposite situation is found in the rural areas.

Table 38 Mobility tool ownership in groups by ARE classification (2005)

Spatial and transport system defined classification	No Car + No Tickets	No Car + HF T	No Car + Nat T / Reg T	Car + No Tickets	Car + HF T	Car + Nat T / Reg T
Main centres	4.4%	9.2%	27.7%	19.4%	20.6%	18.7%
Middle and ancillary centres with railway access	2.2%	6.7%	3.5%	33.3%	36.2%	18.2%
Middle and ancillary centres without railway access	6.0%	5.6%	5.9%	37.3%	30.0%	15.2%
Agglomeration municipalities	4.8%	4.9%	6.0%	36.0%	31.4%	16.9%
Rural areas	3.8%	2.8%	0.7%	50.3%	27.9%	14.6%
Overall	4.6%	6.7%	13.7%	31.0%	26.9%	17.1%

In Table 39 the corresponding binomial logit models for the mobility tool ownership in groups are presented. Persons without any mobility tools tend to be older and foreign nationals as well as have an on average lower income and higher accommodation costs. The group having a car and a national or regional ticket available shows a higher educational degree as well as a higher probability for being in education and in employment. The respondents owning only public transport season tickets are more likely to live in the main centres, whereas car owners are considerably less often found there.

Table 39 Binomial logit models for mobility tool ownership in groups (2005)

Explanatory variable	No Car + No Tickets	No Car + HF T	No Car + Nat T / Reg T	Car + No Tickets	Car + HF T	Car + Nat T / Reg T
Age in years	− 0.214		− 0.448	+ 0.277	+ 0.083	+ 0.113
Age in years squared	+ 0.002		+ 0.003	− 0.002	− 0.001	
Age in years natural logarithm		+ 0.921	+ 9.470	− 4.055		− 5.877
Gender: Male			+ 1.227		− 1.258	
Age in years * Gender: Male			− 0.037	+ 0.020	+ 0.021	
Nationality: Swiss national	− 1.651			− 0.633	+ 1.019	
College or university degree				− 0.852	+ 0.390	+ 0.637
In education		− 1.437	+ 0.662	− 1.654		+ 1.328
In employment						+ 0.934
Monthly income in 1000 CHF	+ 5.046	− 0.127		− 0.250	+ 0.090	
Monthly income in 1000 CHF squared	− 0.842					
Monthly income natural logarithm	− 2.768			+ 0.811		
Accommodation costs in 1000 CHF	− 6.413		+ 4.766	+ 1.161	+ 0.436	− 1.952
Accommodation costs in 1000 CHF squared	+ 1.155		− 0.644	− 0.218		+ 0.207
Accommodation costs natural logarithm	+ 2.678	− 0.588	− 3.858		− 0.745	+ 2.054
Spatial and transport system defined classification: Main centres (referential category)						
Middle and ancillary centres with railway access	+ 3.820	− 1.046	− 2.225	+ 0.915	+ 0.574	+ 1.515
Middle and ancillary centres without railway access	+ 5.303	− 1.419	− 1.822	+ 0.914	+ 0.420	+ 1.619
Agglomeration municipalities	+ 5.578	− 1.379	− 1.702	+ 0.734	+ 0.624	+ 1.498
Rural areas	+ 5.281	− 2.107	− 3.841	+ 1.476	+ 0.396	+ 1.295
Population in the residential municipality in 1000 inhabitants	+ 0.013	− 0.003				+ 0.005
Number of observations	1154	1154	1154	1154	1154	1154
ρ^2 (adjusted)	0.798	0.663	0.512	0.224	0.208	0.410

For all variables: Level of significance ≤ 0.10

Table 40 shows the estimated parameters of a multinomial logit model for the six different alternatives using the group with only a car as reference category. Concerning the persons that do not have any mobility tool at their disposal, age has a negative elasticity of 0.427 for the mean values of the sample. This means that with each percent increase in age the probability for no mobility tools decreases by 0.4%, when all the other explanatory variables remain constant. Regarding the respondents that are older than the sample average, the utility increases with increasing age. From the age of 35 years onwards women show a higher propensity to be in this group. For the other groups, the ownership increases in general with increasing age. Furthermore, males are less likely than females to be in these groups in comparison to the referential group. Respondents with no available mobility tools tend to be foreign nationals, whereas the probability for Swiss persons to own a car and a season ticket at the same time is higher. A college or university degree has a positive effect on the ownership of mobility tools. This also applies to the monthly income with a mean elasticity of 0.961 for the group with national or regional tickets, 0.515 and 0.466 for the last two groups in the table owning a car and public transport season tickets. Only persons not owning any mobility tools show a negative influence for an income higher than 4500 CHF. The costs that are paid for the accommodation influence the probability of being in the referential group positively up to a value of about 2700 CHF. Afterwards, the utility decreases with rising costs. Respondents living in the main centres are the least likely to merely own a car, whereas this is the case more frequently in the rural areas compared to the other groups.

Table 40   Multinomial logit model for mobility tool ownership in groups (2005)

Explanatory variable	No Car + No Tickets	No Car + HF T	No Car + Nat T / Reg T	Car + No Tickets	Car + HF T	Car + Nat T / Reg T
Age in years	− 0.708		− 0.447		− 0.108	− 0.381
Age in years squared	+ 0.006	+ 0.001	+ 0.004		+ 0.001	+ 0.003
Age in years natural logarithm	+ 9.535		+ 6.639		+ 2.474	+ 4.236
Gender: Male	+ 2.315		+ 1.494			
Age in years * Gender: Male	− 0.068	− 0.026	− 0.059		− 0.017	− 0.015
Nationality: Swiss national	− 1.471				+ 1.131	+ 0.438
College or university degree		+ 0.870	+ 0.636		+ 0.873	+ 1.007
In education			+ 1.668		+ 0.990	+ 2.075
In employment						+ 0.723
Monthly income in 1000 CHF	+ 3.070		+ 0.766		+ 0.284	+ 0.237
Monthly income in 1000 CHF squared	− 0.501		− 0.022			
Monthly income natural logarithm	− 2.350		− 1.690		− 0.759	− 0.696

Table 40 is continued ...

Table 40 continued ...

Explanatory variable	No Car + No Tickets	No Car + HF T	No Car + Nat T / Reg T	Car + No Tickets	Car + HF T	Car + Nat T / Reg T
Accommodation costs in 1000 CHF				+ 0.982		
Accommodation costs in 1000 CHF squared				– 0.195		
Accommodation costs natural logarithm				+ 0.141		
Spatial and transport system defined classification: Main centres (referential category)						
Middle and ancillary centres with railway access				+ 0.876		
Middle and ancillary centres without railway access				+ 0.940		
Agglomeration municipalities				+ 0.732		
Rural areas				+ 1.419		
Number of observations						1154
ρ^2 (adjusted)						0.205

For all variables: Level of significance ≤ 0.10

Possibilities to overcome the problems deriving from the independence of irrelevant alternatives (IIA) property of the multinomial logit model include nested logit (NL) and cross-nested logit (CNL) models. Figure 13 shows the structure and Table 41 the estimated parameters of a nested logit model with two nests, one for owning a car and one for not owning a car. Appendix B 1 (Table B 1-1, Table B 1-2, Table B 1-3) presents other possible specifications with two nests. Besides the car and no car option, these include a model with nests for national and regional tickets and no national and regional tickets as well as a model with nests for half-fare discount tickets and no half-fare discount tickets. The model in Table 41 fits the data best. Concerning age, the utility decreases until the age of 52 years and 46 years for the male and female respondents, respectively, those do not have a car or season tickets at their disposal. Afterwards, the utility slowly increases. For all groups, men show in general a lower probability than women in comparison to the persons only owning a car. The Swiss nationality as well as holding a college or university degree has a positive effect on the ownership of mobility tools. This also applies to being in education. Person with no mobility tools tend to have a lower income. For respondents with an available car and public transport season tickets, the utility declines to a minimum value for an income between 2000 CHF and 3000 CHF, and then rises continuously with increasing income. At the same time, accommodation costs influence car ownership positively. With reference to the main centres, respondents living in other spatial and transport system defined types are more likely to merely own a car. The two scale parameters reflect the correlations among the alternatives within each nest which are higher for the alternatives without a car.

Figure 13   Structure of the nested logit model for mobility tool ownership in groups with two nests for car and no car

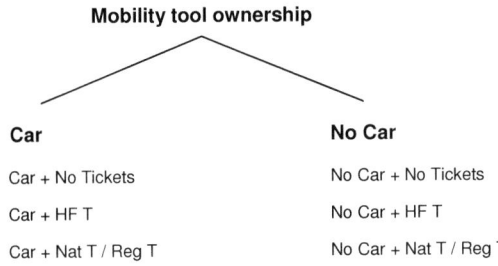

Table 41   Nested logit model for mobility tool ownership in groups with two nests for car and no car (2005)

Explanatory variable	No Car + No Tickets	No Car + HF T	No Car + Nat T / Reg T	Car + No Tickets	Car + HF T	Car + Nat T / Reg T
Age in years	− 0.961		− 0.745			− 1.126
Age in years squared	+ 0.007		+ 0.005			+ 0.008
Age in years natural logarithm	+ 13.788	+ 2.878	+ 13.053		+ 4.120	+ 14.483
Gender: Male		− 2.120			− 2.710	− 2.077
Age in years * Gender: Male	− 0.052		− 0.052			
Nationality: Swiss national		+ 2.287	+ 2.030		+ 5.027	+ 1.971
College or university degree		+ 2.341	+ 1.964		+ 2.876	+ 3.004
In education	+ 3.366	+ 3.487	+ 5.537		+ 4.255	+ 8.077
In employment						+ 1.468
Monthly income in 1000 CHF			+ 1.072		+ 0.673	+ 0.743
Monthly income in 1000 CHF squared			− 0.020			
Monthly income natural logarithm	− 1.592		− 2.365		− 1.371	− 2.263
Accommodation costs natural logarithm				+ 1.456		
Spatial and transport system defined classification: Main centres (referential category)						
Middle and ancillary centres with railway access				+ 3.131		
Middle and ancillary centres without railway access				+ 2.536		
Agglomeration municipalities				+ 2.129		
Rural areas				+ 4.331		

Table 41 is continued ...

Table 41 continued ...

Explanatory variable	No Car + No Tickets	No Car + HF T	No Car + Nat T / Reg T	Car + No Tickets	Car + HF T	Car + Nat T / Reg T
Model parameters for the two nests:						
Nest: Car						0.248 *
Nest: No Car						0.662 *
Number of observations						1154
ρ^2 (adjusted)						0.211

For all variables: Level of significance ≤ 0.10
* For parameters: Level of significance ≤ 0.10

In Figure 14 the structure and in Table 42 the results of a cross-nested logit model with four nests are represented for the ownership of mobility tools in different groups. Concerning age, gender and income, the respondents without any available mobility tools show very similar tendencies as in the nested logit model. For persons only owning season tickets, age has a positive effect, especially for women. At the same time, men are considerably less likely to be in these two groups. This also applies to the respondents with a car and a half-fare discount ticket, whereas the ownership of a car and a national or regional ticket decreases with increasing age and from the age of 33 years onwards men show a higher propensity than women. In comparison to mere car owners, persons having a car and public transport season tickets at their disposal rather tend to be Swiss nationals, tend to hold a college or university degree as well as tend to be in education and employment. In the cases where the monthly income has a significant influence, this influence is positive for the mean value, after a decrease for lower incomes until about 2500 CHF. Concerning the accommodation costs, just the opposite trend is observable for the group of mere car owners. And again, this group is the least likely to live in the main centres. The correlations within the various nests are, with an exception for cars, relatively low. The parameters describing the degree at which an alternative belongs to a nest are primarily not significant. The high value for mere national or regional ticket owners in the corresponding nest indicates that the specification of the model structure is not very appropriate.

Figure 14 Structure of the cross-nested logit model for mobility tool ownership in groups with four nests for car, national and regional tickets, half-fare discount tickets and no mobility tools

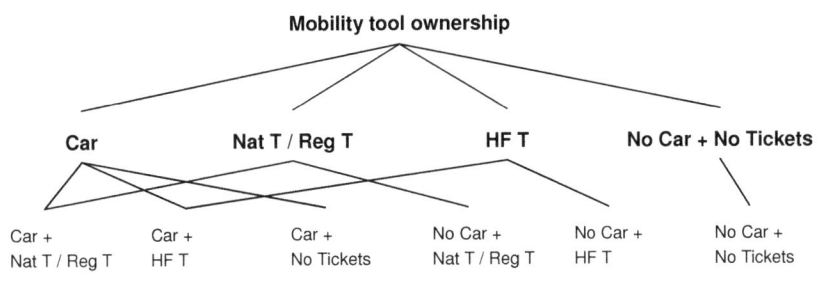

Table 42 Cross-nested logit model for mobility tool ownership in groups with four nests for car, national and regional tickets, half-fare discount tickets and no mobility tools (2005)

Explanatory variable	No Car + No Tickets	No Car + HF T	No Car + Nat T / Reg T	Car + No Tickets	Car + HF T	Car + Nat T / Reg T
Age in years	− 0.398					− 0.157
Age in years squared	+ 0.004	+ 0.001	+ 0.002			
Age in years natural logarithm	+ 2.376				+ 2.369	
Gender: Male	+ 1.631				− 1.728	− 2.441
Age in years * Gender: Male	− 0.066	− 0.035	− 0.077			+ 0.075
Nationality: Swiss national	− 0.789				+ 3.390	+ 2.372
College or university degree		+ 1.170			+ 1.794	+ 2.718
In education					+ 2.359	+ 4.597
In employment						+ 0.958
Monthly income in 1000 CHF			+ 1.176		+ 0.574	
Monthly income in 1000 CHF squared	− 0.110		− 0.020			
Monthly income natural logarithm			− 2.609		− 1.257	
Accommodation costs in 1000 CHF per month			+ 2.021			
Accommodation costs natural logarithm			− 0.385			

Table 42 is continued ...

Table 42 continued ...

Explanatory variable	No Car + No Tickets	No Car + HF T	No Car + Nat T / Reg T	Car + No Tickets	Car + HF T	Car + Nat T / Reg T
Spatial and transport system defined classification:						
Main centres (referential category)						
Middle and ancillary centres with railway access				+ 1.927		
Middle and ancillary centres without railway access				+ 1.832		
Agglomeration municipalities				+ 1.502		
Rural areas				+ 2.872		
Model parameters for the four nests as well as for the six groups:						
Nest: Car						0.345 *
Car + No Tickets						0.002
Car + HF T						0.000
Car + Nat T / Reg T						0.000
Nest: National and regional tickets						0.096 *
No Car + Nat T / Reg T						75.522 *
Car + Nat T / Reg T						0.002
Nest: Half-fare discount tickets						0.041 *
No Car + HF T						0.000
Car + HF T						0.000
Nest: No mobility tools						1.000
No Car + No Tickets						0.133 *
Number of observations						1154
ρ^2 (adjusted)						0.209

For all variables: Level of significance ≤ 0.10
* For parameters: Level of significance ≤ 0.10

Table 43 shows a multivariate probit model for the mobility tool ownership in groups, in which correlations between the different alternatives are incorporated by freeing the variance-covariance matrix. In the model most, but not all of the variables represented are significant at a 0.10 level. The estimated parameters are indicated accordingly. For comparison, Appendix B 1 (Table B 1-4, Table B 1-5) includes two multivariate probit models, one using all variables for all alternatives and one only using all the significant variables of the corresponding multinomial logit model. Compared to the MNL model, there are considerably fewer explanatory variables significant, when the correlations between alternatives are taken into account. In the multivariate probit model shown in Table 43 age and gender do not have a clear effect. For respondents without any mobility tools, the utility decreases until the age of about 32 years, and then increases. In this group more foreign nationals are found. Furthermore, their income tends to be lower. Persons owning a car and season tickets more frequently hold a college or university degree, are employed and have higher incomes. The costs for the accommodation influence mere car ownership negatively. Respondents not living in the main centres are more likely to be in this group. At the same time, the size of the residential municipality has a small positive effect. The correlations between the different alternatives are overall negative, but only in a few cases significant.

Table 43 Multivariate probit model for mobility tool ownership in groups (2005)

Explanatory variable	No Car + No Tickets	No Car + HF T	No Car + Nat T / Reg T	Car + No Tickets	Car + HF T	Car + Nat T / Reg T
Age in years	− 0.029 *	+ 0.051	− 0.053 *			− 0.080 *
Age in years squared	+ 0.000 *	− 0.000	+ 0.001 *			+ 0.001 *
Gender: Male					− 0.778 *	
Age in years * Gender: Male			− 0.007 *		+ 0.012 *	− 0.000
Nationality: Swiss national	− 0.828 *					
College or university degree					+ 0.243 *	+ 0.313 *
In education					− 0.276 *	+ 0.788 *
In employment					+ 0.057	+ 0.399 *
Monthly income in 1000 CHF	− 0.254 *	+ 0.270 *				
Monthly income in 1000 CHF squared		− 0.019 *			+ 0.005 *	
Monthly income natural logarithm		− 0.448 *			− 0.102 *	+ 0.064
Accommodation costs natural logarithm				− 0.111 *		
Spatial and transport system defined classification: Main centres (referential category)						
Middle and ancillary centres with railway access				+ 0.374		
Middle and ancillary centres without railway access				+ 0.484 *		
Agglomeration municipalities				+ 0.444		
Rural areas				+ 0.800 *		
Population in the residential municipality in 1000 inhabitants				+ 0.000		
Correlation matrix:						
No Car + No Tickets	+ 1.000	− 0.054	− 0.085	− 0.157 *	− 0.129 *	− 0.106
No Car + HF T		+ 1.000	− 0.090	− 0.172	− 0.138 *	− 0.118
No Car + Nat T / Reg T			+ 1.000	− 0.268	− 0.220	− 0.184
Car + No Tickets				+ 1.000	− 0.410 *	− 0.336
Car + HF T					+ 1.000	− 0.280
Car + Nat T / Reg T						+ 1.000
Number of observations						1154
ρ^2 (adjusted)						0.503

* Level of significance ≤ 0.10

In order to compare the various discrete choice models, the log-likelihood ratio test is applied, which is possible, when the same choice variable is used (Hensher, Rose and Greene, 2005). In this context, the final log-likelihoods of two models are taken into account, calculating the log-likelihood ratio G^2 for testing the null hypothesis that the one model is not better than the other model as follows

$$G^2 = -2(L(largest) - L(smallest)), \tag{7-2}$$

where $L(largest)$ and $L(smallest)$ are the largest and the smallest log-likelihood of the two models under comparison, respectively. This ratio is compared to the corresponding χ^2 statistic. Thereby, the degree of freedom is equal to the difference between the numbers of parameters estimated in the two models.

In Table 44 the various discrete choice models for the mobility tool ownership in six different groups are described, specifying the initial and final log-likelihood as well as the number of estimated parameters. Besides the summarised binomial logit models, the logit models are relatively similar to one another, with the best model being the nested logit model with the two nests for owning a car and not owning a car. Concerning the different multivariate probit models, the model only using the variables that are significant at a 0.10 level shows the highest predictive capability. The other two probit models do not significantly improve the log-likelihood over the first model.

Table 44 Description of the various models for mobility tool ownership in groups (2005)

Models for mobility tool ownership in groups	Initial log-likelihood	Final log-likelihood	Number of parameters	ρ^2 (adjusted)
Binomial logit models	−4800.7	−2471.8	77	0.469
Multinomial logit model	−2068.3	−1596.1	48	0.205
Nested logit model with two nests for car and no car	−2068.3	−1586.9	45	0.211
Nested logit model with two nests for national and regional tickets and no national and regional tickets	−2068.3	−1590.0	44	0.210
Nested logit model with two nests for half-fare discount tickets and no half-fare discount tickets	−2068.3	−1595.7	46	0.206
Cross-nested logit with four nests for car, national and regional tickets, half-fare discount tickets and no mobility tools	−2068.3	−1589.6	47	0.209
Multivariate probit model	−4803.5	−2340.5	47	0.503
Multivariate probit model using all variables for all alternatives	−4803.5	−2302.5	84	0.503
Multivariate probit model using the variables of the corresponding multinomial logit model	−4803.5	−2447.4	63	0.477

Regarding the significant explanatory variables and the estimated parameters, the various logit models also strongly resemble one another. The most important variables, occurring in all the models, are age, gender, nationality, holding a college or university degree, income and accommodation costs as well as the classification of the residential municipality. Between the logit models and the probit model, larger differences occur. In the probit model, considerably fewer variables are significant. Their impact is captured by the correlations between the different mobility tool ownership groups that are incorporated in the probit model.

8 Analyses for the period from 1985 to 2004

In this chapter, the data for the period of the 20 years from 1985 to 2004 is analysed, which was collected in a retrospective survey by means of a multidimensional life course calendar. Overall, 1140 of the 1166 persons participating in the survey filled in the life course calendar, which amounts to a share of 97.8%. A general description is followed by an exploration of the places of residence and the moving behaviour on one side as well as of the ownership of mobility tools on the other side. Analyses over time, over the life course and concerning the birth cohort membership of respondents as well as analyses of various durations and occurring changes are carried out.

8.1 General description

Figure 15 and Figure 16 show the occurrence of important personal and familial events during the observed time period from 1985 to 2004 and during the life course, respectively. In regard to age, five year intervals are used. Concerning the 20 year time period, the shares of the different events range between 0% and 2% with the lowest share for break-ups and divorces. Overall, no large variation is observable. The move out of the parents' house primarily takes place at the age between 15 and 20 years. The birth of persons in the household shows a maximum at the beginning, covering the respondents' birth as well as the birth of siblings. Between the ages of 20 and 30 years again a higher number of children are born, followed by a gradual decrease afterwards. The share for the formation of partnerships and marriages is highest for persons aged from 15 to 30 years. Break-ups and divorces show overall a relatively flat graph with values not exceeding 1%.

Appendix B 2 (Figure B 2-1) illustrates the personal and familial events by gender, age and birth cohort membership. Overall, the same tendencies are noticeable, but the curves are not as smooth due to the lower number of observations in each of the various groups.

Figure 15 Personal and familial events by time (1985-2004)

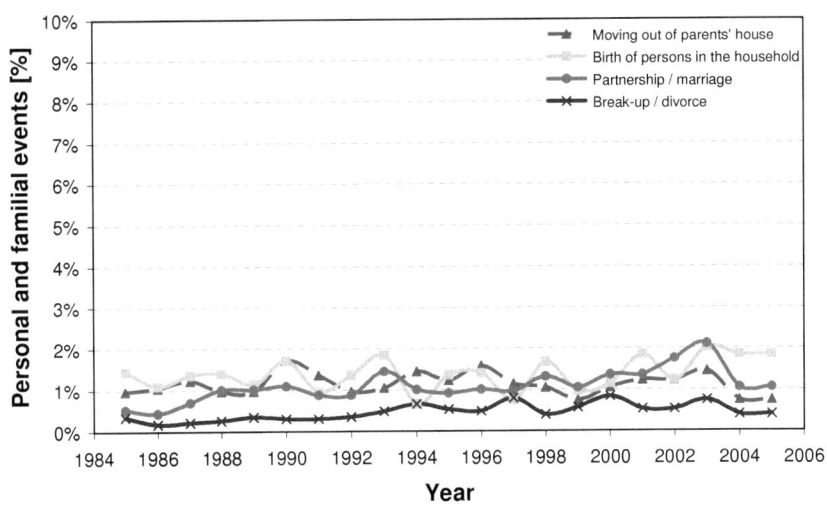

Figure 16 Personal and familial events by age (1985-2004)

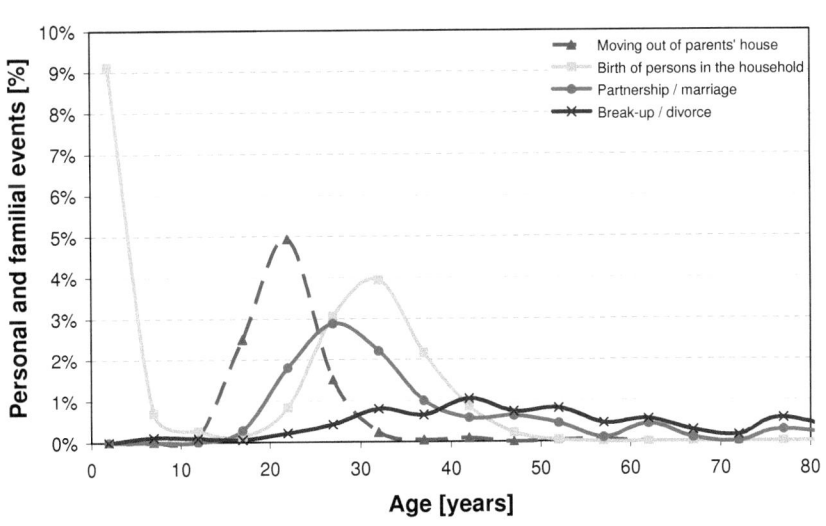

8.2 Places of residence and moving behaviour

Figure 17 illustrates the distribution of the residential durations for the period from 1985 to 2004, including both uncensored and censored durations. Overall, 4155 durations are observed within this period. On average these durations are about 7 years long with a standard deviation of about 9 years. Approximately two thirds of the durations are up to five years, nearly 80% up to ten years and nearly 90% up to twenty years long.

In Table 45 the mean values and the standard deviations as well as the 25%-, 50%- and 75%-percentiles of the residential durations are shown for the different spatial and transport system defined municipality types. The five types vary significantly from one another. In the middle and ancillary centres with access to the national railway network persons tend to move most frequently, whereas in the centres without such an access the residential durations are the longest ones.

Figure 17 Distribution of the residential durations (1985-2004)

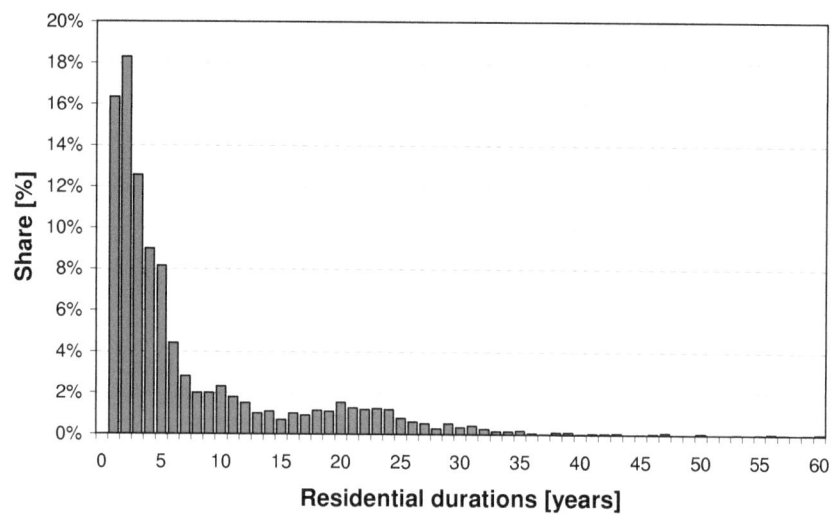

Table 45   Residential durations by ARE classification (1985-2004)

Spatial and transport system defined classification	Mean value	Standard deviation	25%-Percentile	50%-Percentile	75%-Percentile
Main centres	7.2 years	9.7 years	2.0 years	3.5 years	7.5 years
Middle and ancillary centres with railway access	6.0 years	6.6 years	1.5 years	3.0 years	8.1 years
Middle and ancillary centres without railway access	8.3 years	9.0 years	2.5 years	4.5 years	11.0 years
Agglomeration municipalities	7.0 years	8.2 years	2.0 years	3.5 years	8.0 years
Rural areas	7.5 years	7.9 years	1.5 years	4.0 years	11.0 years
Abroad	7.0 years	8.3 years	1.0 years	3.5 years	9.5 years
Overall	7.1 years	8.6 years	1.5 years	3.5 years	9.0 years

The values in parentheses indicate the standard deviations of the mean values.

Figure 18 presents the distribution of the distances between two successive places of residence, again showing the mean value, the standard deviation as well as the 25%-, 50%- and 75%-percentiles. The distance to the previous place of residence is only known for 2436 moves, since this information is always missing for the first observed place of residence within the period from 1985 to 2004. Overall, a place of residence is approximately 481 kilometres away from the previous one with a standard deviation of 2109 kilometres. The distribution of the residential distances is very strongly left-skewed. Over one third of all the moves take place within a radius of 5 kilometres. The shown range from 0 kilometres to 250 kilometres covers about 90% of all residential distances. The maximum value lies around 19339 kilometres. This confirms the statement that most residential moves are characterised by short distances (Blijie, 2005; Franz, 1984).

Table 46 describes the distances between two places of residence by the ARE classification. The residential distances are also significant different for the five municipality types. In this context, the observed distances are smallest in the rural areas and by contrast considerably above average, when persons move abroad, consistent with the expectations.

Figure 18 Distribution of the residential distances (1985-2004)

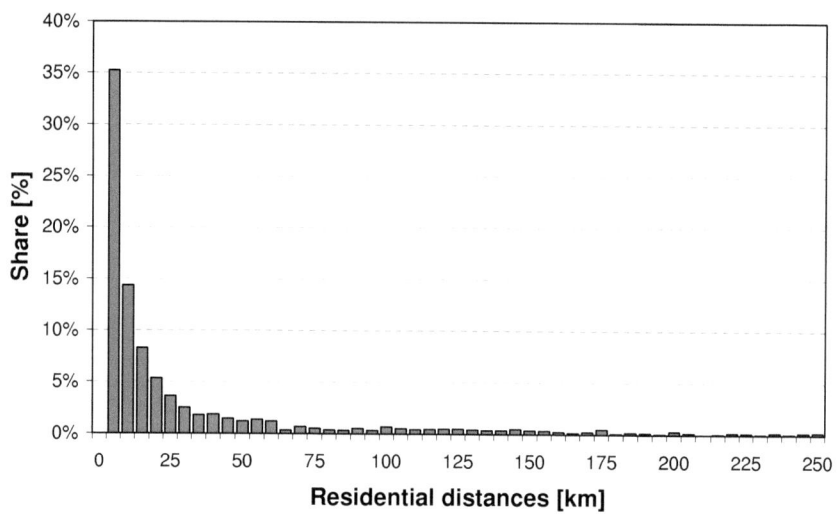

Table 46 Residential distances by ARE classification (1985-2004)

Spatial and transport system defined classification	Mean value	Standard deviation	25%-Percentile	50%-Percentile	75%-Percentile
Main centres	383.9 km	1834.3 km	1.8 km	7.6 km	48.5 km
Middle and ancillary centres with railway access	222.7 km	1378.5 km	2.5 km	12.5 km	37.3 km
Middle and ancillary centres without railway access	443.4 km	2185.0 km	2.2 km	7.7 km	26.3 km
Agglomeration municipalities	226.5 km	1445.2 km	3.1 km	9.1 km	25.4 km
Rural areas	159.3 km	936.7 km	4.0 km	17.7 km	49.1 km
Abroad	2730.4 km	4482.5 km	3.1 km	434.0 km	2620.3 km
Overall	481.3 km	2109.1 km	2.5 km	10.2 km	47.2 km

The values in parentheses indicate the standard deviations of the mean values.

Table 47 shows the moving directions in regard to the ARE classification during the observed 20 year period from 1985 to 2004. The proportion of unknown places of residence is relatively high. This is again primarily connected to the fact that the previous place of residence for the first one observed within this period is not known. Overall, moves within one municipality type account for about one third of all moves. The highest value for each type is found in the main diagonal, where the observed counts are considerably higher than the expected counts. There exists a significant relation with a positive association between the spatial and transport system defined type of the residential municipality before and after a move.

Table 47   Directions of all moves by ARE classification (1985-2004)

Previous place of residence Place of residence	Type 1	Type 2	Type 3	Type 4	Type 5	Abroad	Un-known	Overall
Main centres	9.1%	1.0%	1.1%	2.8%	1.7%	1.7%	8.1%	25.4%
Middle and ancillary centres with railway access	0.8%	1.7%	0.4%	1.4%	0.8%	0.2%	1.8%	7.1%
Middle and ancillary centres without railway access	1.9%	0.2%	4.1%	2.5%	0.7%	0.8%	4.4%	14.7%
Agglomeration municipalities	2.8%	1.1%	2.6%	7.1%	1.3%	0.7%	8.1%	23.7%
Rural areas	0.7%	0.7%	0.5%	1.5%	3.1%	0.4%	3.7%	10.5%
Abroad	0.5%	0.2%	0.3%	0.5%	0.1%	2.7%	3.0%	7.3%
Unknown	1.2%	0.3%	0.2%	0.8%	0.4%	0.8%	7.8%	11.4%
Overall	17.0%	5.1%	9.1%	16.5%	8.0%	7.2%	37.0%	100.0%

Table 48 compares the accommodation size indicated by the number of rooms of successive places of residence. The respondents tend to move to slightly bigger accommodations. This especially applies for the smaller accommodations with sizes ranging from one room to three rooms. Along the main diagonal and the adjacent fields the observed counts are to some extent higher than the expected counts, whereas the standardised residuals are negative for the other combinations. Overall, the number of rooms before and after a move is positively correlated in a significant way.

Table 48 Directions of all moves by accommodation size (1985-2004)

Previous place of residence Place of residence	One room	Two rooms	Three rooms	Four rooms	Five rooms	Six rooms +	Un-known	Overall
One-room-accommodation	0.7%	0.5%	0.6%	0.6%	0.4%	0.8%	1.2%	4.9%
Two-room-accommodation	0.9%	1.0%	1.3%	1.1%	1.0%	0.9%	2.1%	8.3%
Three-room-accommodation	1.0%	1.9%	3.6%	3.0%	1.8%	2.0%	5.2%	18.5%
Four-room-accommodation	0.7%	1.6%	3.8%	3.9%	2.6%	2.0%	7.7%	22.3%
Five-room-accommodation	0.4%	0.7%	2.8%	3.9%	3.3%	1.8%	6.6%	19.5%
Six-room-accommodation and more rooms	0.2%	0.6%	1.0%	2.7%	2.5%	1.6%	6.4%	14.9%
Unknown	0.4%	0.4%	0.9%	0.6%	0.7%	0.6%	7.9%	11.5%
Overall	4.4%	6.7%	14.1%	15.8%	12.3%	9.7%	37.0%	100.0%

Table 49 shows the reasons for moving indicated by the respondents, who had the possibility to specify multiple answers. In the first place personal and familial reasons with 41% are given, followed by accommodation related reasons with 26%. However, these two categories are closely connected to one another. In general, motives behind residential mobility are difficult to separate, and not easily to assign. In many cases several factors together lead to a move (Birg and Flöthmann, 1992). Education and employment related reasons have a share of about 26%. This reason is especially important for moving abroad, whereas other categories play a subordinate role for this kind of move. Overall, the shares vary significantly from one another in regard to the ARE municipality types for all reasons of moving.

Table 49 Reasons for all moves by ARE classification (1985-2004)

Spatial and transport system defined classification	Personal and familial reasons	Education and employment related reasons	Accommodation related reasons	Surrounding related reasons	Vicinity to family and friends	Other reasons
Main centres	41.7%	29.1%	33.2%	17.3%	7.9%	1.8%
Middle and ancillary centres with railway access	46.3%	33.0%	21.8%	12.2%	9.5%	2.7%
Middle and ancillary centres without railway access	42.0%	31.5%	35.1%	12.8%	11.5%	2.1%
Agglomeration municipalities	53.5%	22.1%	32.3%	13.7%	11.6%	0.8%
Rural areas	50.5%	27.6%	21.9%	14.2%	10.0%	0.9%
Abroad	38.4%	44.4%	14.6%	4.6%	3.6%	3.6%
Overall	41.1%	25.8%	26.2%	12.3%	8.5%	1.6%

Figure 19 and Figure 20 illustrate the share of respondents that are engaged in education and employment as well as their monthly income, once by time and once by age, respectively. At the beginning of the observed period nearly one third of the persons are in education and employment. Over time persons in education diminish till approximately 9% in the year 2005, whereas the share of employed persons increases up to 72%. Accordingly, the income per month, without considering inflation, rises as well by 87% from a level of about 3100 CHF in the year 1985 to a level of about 5700 CHF in the year 2005. With regard to the age of the respondents, the share of persons in education reaches a maximum at the age of about 12 years. Afterwards, it strongly decreases until the age of about 32 years. Employment shows a clear increase between the ages of 15 and 30 years, followed by a relative stable period with a share of about 80% until persons start retiring, when they reach the age of circa 60 years. The monthly income continuously increases over the life course, especially for persons aged from 15 to 30 years. Only after the age of 65 years a small reduction is observable.

Figure 19   Persons in education and employment as well as the monthly income by time (1985-2004)

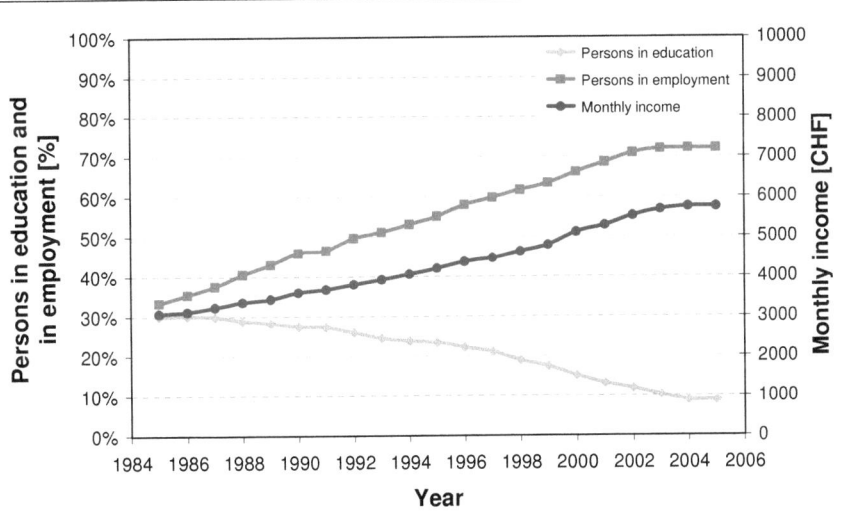

Figure 20 Persons in education and employment as well as the monthly income by age (1985-2004)

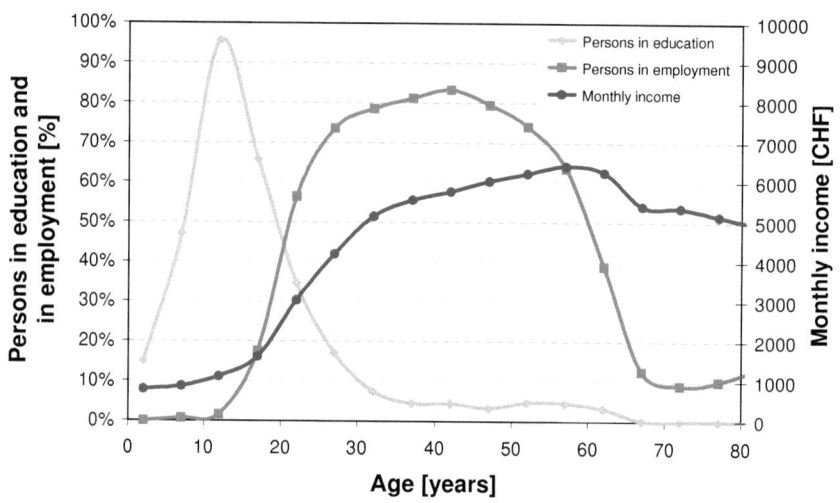

Table 50 and Table 51 describe the air-line distances to education and employment, respectively, by the spatial and transport system defined classification, showing the mean value, the standard deviation as well as the median and the 25%- and 75%-percentiles. On average the place of education is 27 kilometres and the place of employment 21 kilometres away. The values for the different municipality types vary significantly. Abroad the distances are considerably longer, especially for employment, followed by the rural areas. Concerning the places of residence located abroad, the given percentiles with the value zero indicate that the resolution at municipality level is relatively imprecise, as there is no detailed calculation and differentiation within municipalities possible. This means that for persons living and being occupied in the same municipality the distance to their place of education and employment is set to zero kilometres.

Table 50   Distance to the places of education by ARE classification (1985-2004)

Spatial and transport system defined classification	Mean value	Standard deviation	25%-Percentile	50%-Percentile	75%-Percentile
Main centres	12.8 km	52.3 km	1.4 km	2.6 km	5.9 km
Middle and ancillary centres with railway access	28.6 km	358.5 km	1.3 km	2.2 km	16.5 km
Middle and ancillary centres without railway access	12.7 km	49.6 km	1.5 km	6.8 km	11.8 km
Agglomeration municipalities	11.9 km	37.5 km	1.1 km	4.8 km	12.1 km
Rural areas	46.4 km	560.5 km	0.8 km	5.5 km	16.7 km
Abroad	85.1 km	469.5 km	0.0 km	0.0 km	0.0 km
Overall	27.2 km	294.0 km	0.9 km	3.5 km	11.2 km

Table 51   Distance to the places of employment by ARE classification (1985-2004)

Spatial and transport system defined classification	Mean value	Standard deviation	25%-Percentile	50%-Percentile	75%-Percentile
Main centres	22.1 km	345.8 km	1.8 km	3.8 km	6.6 km
Middle and ancillary centres with railway access	15.1 km	24.1 km	1.6 km	5.6 km	16.9 km
Middle and ancillary centres without railway access	10.2 km	26.2 km	1.9 km	6.3 km	10.8 km
Agglomeration municipalities	12.3 km	20.9 km	3.6 km	7.9 km	13.1 km
Rural areas	23.9 km	240.3 km	1.3 km	8.2 km	19.0 km
Abroad	142.8 km	767.8 km	0.0 km	0.0 km	21.4 km
Overall	21.3 km	254.0 km	1.7 km	5.5 km	11.9 km

Figure 21 and Figure 22 show the median distances between the place of residence and the corresponding places of education and employment for the observed time period from 1985 to 2004 and for the age of the respondents, respectively. The distances to both locations increase over time, to a greater extent for education than for employment. In regard to age the distance to the place of education shows a strong rise between the ages of 10 and 18 years. With increasing age the distance to education fluctuates, which is connected to more specialised educations and a rather low number of observed cases for the older age groups. For the distance to the place of employment, there is likewise a strong increase noticeable until the age of 18 years. Then, a slow decrease until the retirement age and an afterwards stable section with a median distance of about 1 kilometre follow. Overall, the median values are substantially lower than the mean values.

Figure 21   Median distances to the places of education and employment by time (1985-2004)

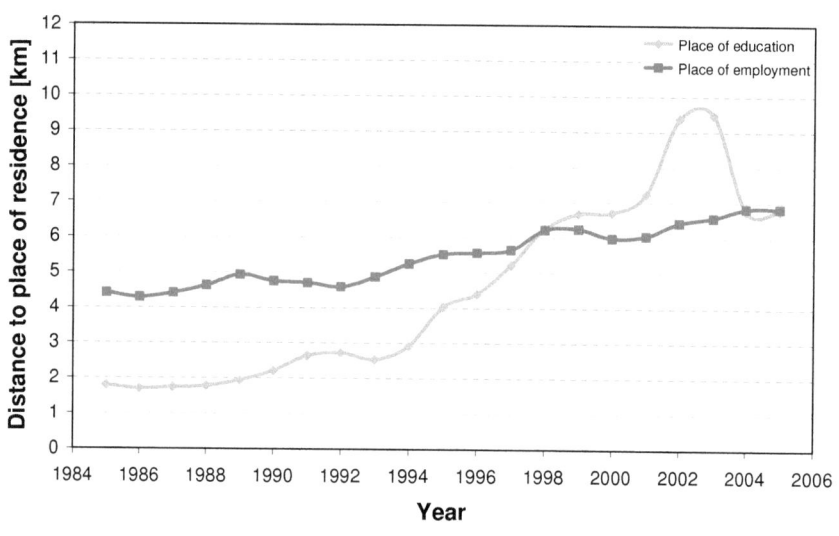

Figure 22   Median distances to the places of education and employment by age (1985-2004)

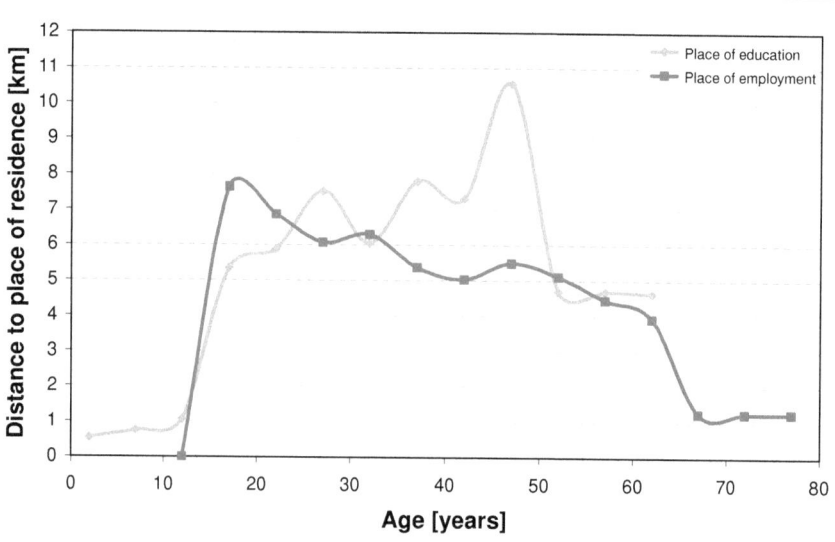

In the following analyses, only respondents in education and employment are considered. Figure 23 and Figure 24 show the mode of transport most frequently used for the trip from home to the places of education and employment in regard to time and age. The white area up to the full 100% represents the missing data. During the observed 20 year period major changes concerning the main mode of transport to the place of education occur. There is a small increase in the share of private transport by about 10% and a much bigger increase in the share of public transport by over 30%, whereas cycling and walking strongly decline. By contrast, the main mode for the trip to the place of employment remains relatively stable over time. Approximately 50% of the respondents use private transport, 30% public transport and nearly 10% each cycle and walk. Over the life course similar trends are observable. For persons in education, there is a strong increase in the usage of private and public transport, whereas walking is only until the age of about twelve years of some importance. For the trip to the place of employment, persons aged from 25 to 65 years show a relative stable modal split. These developments are closely connected to the changes in the median distance from home to the places of education and employment over time and during the life course, as shown in Figure 21 and Figure 22.

Figure 23   Mode of transport to education and employment by time (1985-2004)

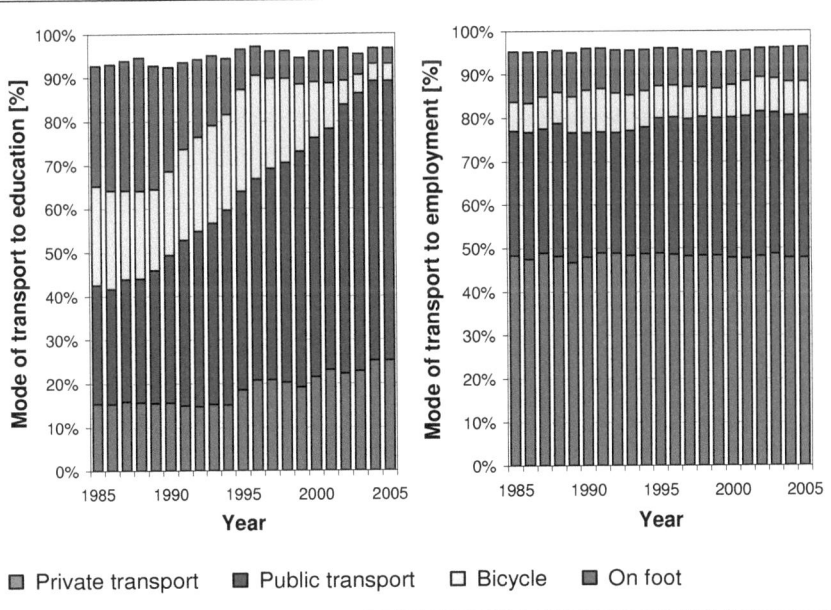

Figure 24   Mode of transport to education and employment by age (1985-2004)

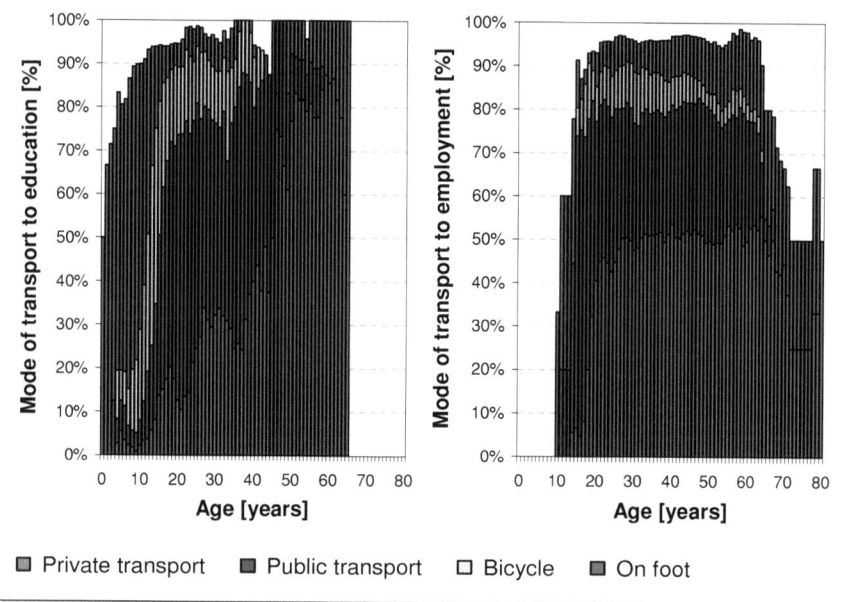

☐ Private transport   ■ Public transport   ☐ Bicycle   ■ On foot

8.3   Ownership of mobility tools

The mobility tools considered in the retrospective survey are cars and different public transport season tickets, including national annual tickets, regional annual and monthly tickets as well as half-fare discount tickets. In Table 52 the mobility tool ownership of all persons is described for the period from 1985 to 2004 by the ARE type of the residential municipality. Overall, 48% of the respondents have a car always and 11% partially at their disposal. Concerning public transport, 7% and 17% own a national or a regional season ticket, respectively, and 31% a half-fare discount ticket. For all mobility tools, significant differences between the spatial and transport system defined municipality types arise. Car availability and public transport season ticket ownership are lowest abroad. In Switzerland, the main centres also show low car ownership, whereas the share of season ticket owners is comparatively high, which is concurrent with the expectations and the literature (Beige, 2004; Karlaftis and Golias, 2002).

Table 52   Mobility tool ownership by ARE classification (1985-2004)

Spatial and transport system defined classification	Car availability: Always	Car availability: Partially	National ticket ownership	Regional ticket ownership	Half-fare discount ticket ownership
Main centres	40.8%	14.2%	9.8%	29.0%	41.8%
Middle and ancillary centres with railway access	52.7%	12.1%	10.1%	11.1%	36.7%
Middle and ancillary centres without railway access	63.1%	10.6%	5.4%	15.6%	36.7%
Agglomeration municipalities	56.2%	11.5%	4.1%	17.1%	32.5%
Rural areas	50.2%	13.2%	6.3%	10.0%	27.8%
Abroad	31.7%	11.0%	0.5%	8.0%	6.9%
Overall	47.9%	11.1%	6.5%	17.3%	31.3%

In Figure 25 and Figure 26 the ownership of mobility tools is shown for all respondents by time and age, respectively. The denoted abbreviations stand for national tickets (Nat T), regional tickets (Reg T) and half-fare discount tickets (HF T). During the 20 year period from 1985 to 2004 the ownership of all mobility tools increases. The availability of only a car declines over time, whereas the share of car and public transport season ticket owners increases from 20% to 45%. At the same time, respondents without any mobility tools diminish during the observed time period. Regarding the age of the respondents, there is, as expected, a strong increase in car ownership after reaching the age of 18 years. Persons aged from 25 to 50 years show the highest share with about 75%. Then, a slow decrease is visible. The ownership of national tickets increases over the life course, whereas the share of regional tickets decreases. The half-fare discount tickets have growing shares. About one third of the respondents own a car and public transport season tickets at the same time. Overall, the ownership of mobility tools increases at the beginning and then remains relatively stable over the life course with only approximately 10% of persons not having any mobility tool at their disposal.

Figure 25 Mobility tool ownership by time (1985-2004)

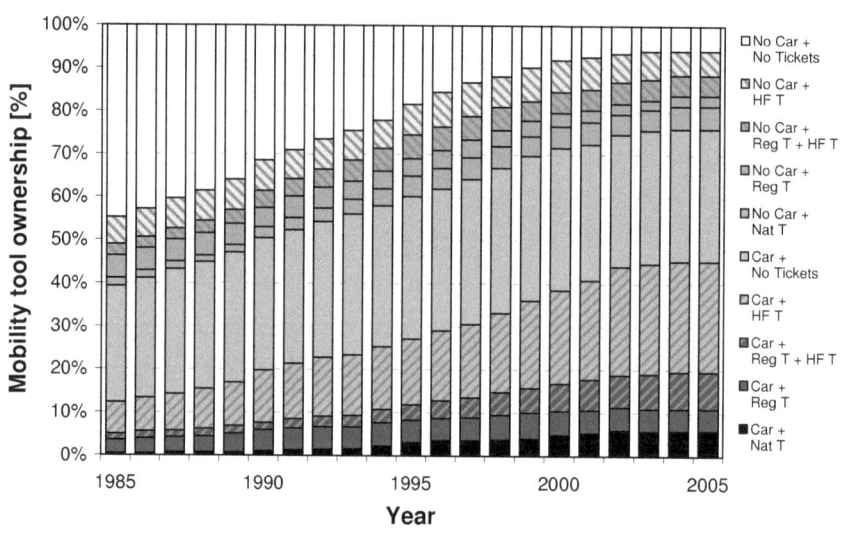

Figure 26 Mobility tool ownership by age (1985-2004)

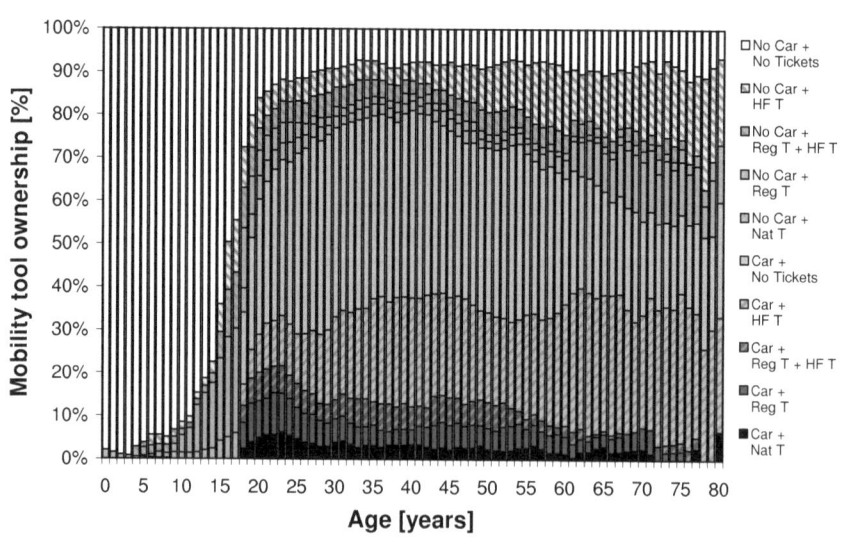

Below, the ownership of mobility tools is analysed by age of the respondents and their membership in birth cohorts. Thereby, it is possible to take into account changes during the life course of a person as well as cohort effects which specify intra- and inter-generational similarities and varieties in one generation or between different generations (Armoogum, Madre and Bussière, 2002; Ryder, 1965). In this context, it is assumed that people born in the same time interval and, therefore, ageing together also share a common life experience, due to the fact that general changes have differing impacts for persons of unlike age and that the consequences of these changes persist in the subsequent behaviour of these individuals and, thus, of their cohorts (Ryder, 1965). A third temporal dimension includes period effects indicating the impact of the global context and which are independent from the persons (Armoogum et al., 2002; Mayer and Huinink, 1990). Figure 27 illustrates the ownership of cars, including always and partially available cars, by gender, age and birth cohort membership. For all three variables, significant differences occur. The oldest cohort group comprises of people which are born before 1930. The following cohorts span a period of ten years in each case. It is noticeable that the oldest cohort group owns considerably fewer cars than the younger cohorts. Highest is the ownership among those who are 35 to 55 years old today. At the same time, men have noticeably more frequently a car at their disposal than women of the same age. However, for the younger generations this difference diminishes. Except for the oldest group, there are still increases in ownership observable within the different cohorts. This means that the level of saturation is not necessarily reached yet. The same patterns are observed for Switzerland (Beige, 2004) and the United Kingdom (Dargay, 2001) using national survey data and the so-called pseudo-panel approach, in which panel data is constructed by tracing cohorts in the cross-sectional data and treating the cohort averages for each point in time as observations (Dargay, 2001). Dargay (2001) argues that the age and cohort effects are largely explained by differences in income over the life course and differences in income between generations.

In Figure 28 the ownership of national and regional season tickets is likewise shown by gender, age and birth cohort membership. Compared to car ownership, there is a different trend visible. The cohorts with a current age between 35 and 55 years now show the lowest ownership rates. Furthermore, women generally own more public transport season tickets than men. Therefore, both car ownership on the one side and national and regional season ticket ownership on the other side substitute one another. This is consistent with the findings of Simma and Axhausen (2003).

Figure 29 describes the ownership of half-fare discount tickets separate for men and women, age and birth cohort membership. Over the life course a relatively strong increase in ownership is observed. With the exception of the oldest cohort group, the female respondents tend to own more half-fare discount tickets than the male respondents of the same age.

Figure 27 Car ownership by gender, age and birth cohort membership (1985-2004)

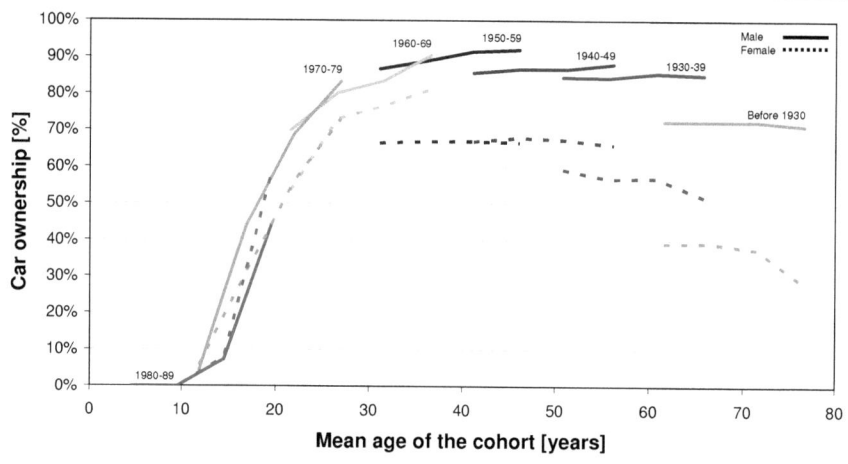

Figure 28 National and regional ticket ownership by gender, age and birth cohort membership (1985-2004)

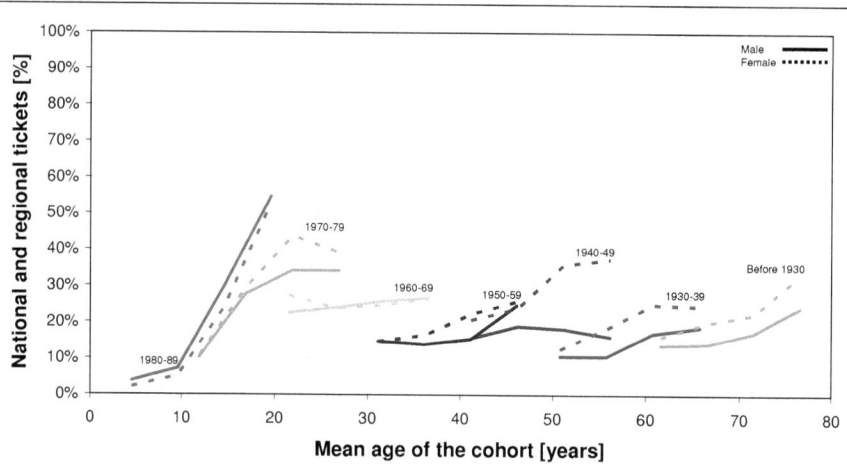

Figure 29 Half-fare discount ticket ownership by gender, age and birth cohort membership (1985-2004)

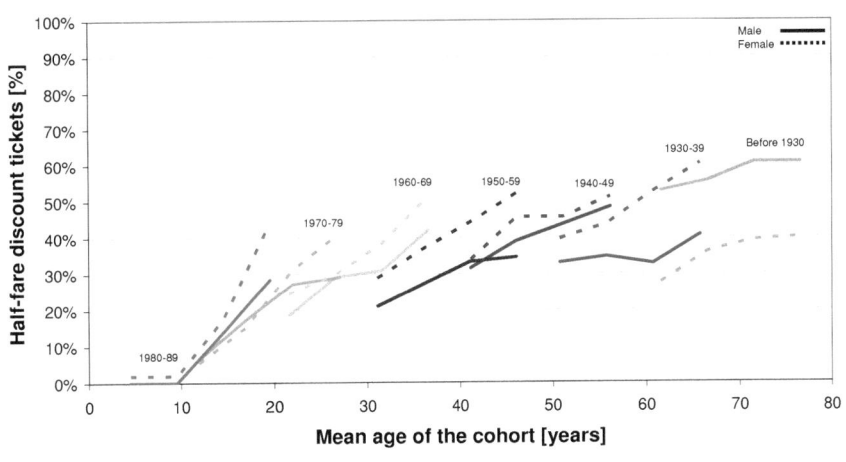

Discrete choice modelling for the ownership of mobility tools

In the following, various univariate and multivariate discrete choice models are estimated for the ownership of mobility tools between 1985 and 2004. For this time period, observations on a semi-annual basis are included. Only persons that are 18 years and older are considered. Overall, 28808 observations are in the data set. Table 53 shows the explanatory variables with their mean values, standard deviations, minima and maxima as well as the source of the data.

Table 53 Description of the explanatory variables (1985-2004)

Explanatory variable	Mean value	Standard deviation	Minimum	Maximum	Source
Age in years	36.137	15.609	0.000	90.500	Retrospective survey (2005)
Gender: Male	0.517	0.500	0.000	1.000	Retrospective survey (2005)
Nationality: Swiss national	0.862	0.345	0.000	1.000	Retrospective survey (2005)
College or university degree	0.341	0.474	0.000	1.000	Retrospective survey (2005)

Table 53 is continued ...

Table 53 continued ...

Explanatory variable	Mean value	Standard deviation	Minimum	Maximum	Source
In education	0.195	0.396	0.000	1.000	Retrospective survey (2005)
In employment	0.650	0.477	0.000	1.000	Retrospective survey (2005)
Monthly income in 1000 CHF	4.718	3.276	0.006	16.000	Retrospective survey (2005)
Car availability: Always	0.573	0.495	0.000	1.000	
Car availability: Partially	0.137	0.344	0.000	1.000	
National ticket ownership	0.076	0.264	0.000	1.000	Retrospective survey (2005)
Regional ticket ownership	0.207	0.405	0.000	1.000	
Half-fare discount ticket ownership	0.394	0.489	0.000	1.000	
Moving out of parents' house	0.010	0.100	0.000	1.000	Retrospective survey (2005)
Birth of a person in the household	0.015	0.120	0.000	1.000	Retrospective survey (2005)
Number of persons in the household	2.752	1.341	1.000	30.000	Retrospective survey (2005)
Number of rooms in the accommodation	4.452	1.660	1.000	30.000	Retrospective survey (2005)
Degree of urbanisation:					Statistical Office of the European Communities (2006)
Urban (referential category)	0.370	0.483	0.000	1.000	
Urban to rural	0.508	0.500	0.000	1.000	
Rural	0.122	0.327	0.000	1.000	
Place of residence abroad	0.024	0.152	0.000	1.000	Retrospective survey (2005)
Purchasing power index in the residential region *	92.583	14.658	59.000	141.000	Statistical Office of the European Communities (2006)

* The index of purchasing power is a combined index of consumer prices and exchange rates, aiming to account for variation in consumer prices and the movement of currency exchange rates over time. It measures the changes in consumer prices in a country in Euro, making an adjustment for changes in exchange rates (Ascoli, 2000; Olsson, 2005).

In the context of the model estimation, it is necessary to take into account that each respondent appears several times as observation and, therefore, to control for unobserved characteristics of the individuals. Thus, an error term is added, which allows individuals who are homogeneous in their observed characteristics to be heterogeneous in their response probabilities (Hsiao, 2003). Within the model specification a random parameter is introduced, which is normally distributed across the entire sample, but invariant for each individual. For this parameter, the standard deviation is estimated, while the mean value is set to zero (Bierlaire, 2005). In Table 54 the results of different binomial logit models for the availability of cars and the ownership of public transport season tickets during the 20 year period are represented. All relevant explanatory variables are used and a corresponding indication of their significance is shown, since estimating these models requires a lot of computational time and effort. The probability of disposing of a car which is available at all times increases until the age of 54 years and then slowly declines. At the same time, male respondents are more often in this position than female respondents. Employment as well as income has a positive influence. The ownership of season tickets is related to a lower proportion of always available cars. In larger households these cars are less likely, as available cars are more likely to be shared. With reference to the urban areas, persons having their own car live more often in rural areas. Rather opposed tendencies are noticeable for partially available cars. For instance, men tend to have a car less frequently only part time at their disposal than women. Furthermore, the monthly income has a negative effect. Persons owning public transport season tickets are in general more likely to simultaneously have a partially available car. The ownership of national and regional tickets for public transport is reduced with increasing age, while the utility for half-fare discount tickets also increases. Men tend to own less public transport season tickets than women. The Swiss nationality as well as a college or university degree and being in education lead to a higher ownership of these mobility tools. Only for the national tickets the distance to the place of employment has a positive influence. Concurrent with the expectations, simultaneous car availability decreases the ownership of public transport season tickets. Persons with national or regional tickets live more often in urban areas. In the cases where the place of residence is abroad, season ticket ownership tends to be lower. The index of purchasing power in the region of residence has a positive influence on the public transport season ticket ownership. All of these results are in general consistent with other analyses concerning the ownership of mobility tools (Beige, 2004; Karlaftis and Golias, 2002; Simma and Axhausen, 2003).

The high values for the standard deviation σ of the individual-specific error term indicate a substantial heterogeneity in the sample (Hsiao, 2003). The sign of the random parameter is not relevant. Appendix B 2 (Table B 2-1, Table B 2-2) shows a comparison of the different binomial logit models for car availability and public transport season ticket ownership, on the one hand not taking the panel effect into account and on the other hand taking the panel effect into account. Overall, the absolute values of the estimated parameters tend to be to some extent larger in the second case, which is also shown in Table 54. The individual-specific random parameters are in all five models significant and play an important role, as the strong increase of the measures for the goodness of fit indicates.

Table 54 Binomial logit models for car availability and public transport season ticket ownership (1985-2004)

Explanatory variable	Car availability: Always	Car availability: Partially	National ticket ownership	Regional ticket ownership	Half-fare discount ticket ownership
Age in years	+ 0.538 *	− 0.089	− 0.474 *	− 0.140 *	+ 0.128 *
Age in years squared	− 0.005 *	− 0.000	+ 0.006 *	+ 0.001	− 0.001
Gender: Male	+ 3.595 *	− 0.842 *	+ 0.972 *	− 0.114	− 1.710 *
Nationality: Swiss national	− 0.147	+ 1.724 *	+ 4.309 *	− 0.053	+ 3.560 *
College or university degree	− 0.245	+ 2.021 *	+ 2.848 *	+ 1.609 *	+ 2.590 *
In education	+ 0.000	− 0.757 *	+ 0.708 *	+ 1.119 *	+ 0.562 *
Change in education	− 0.340 *	+ 0.096	+ 0.186	+ 0.138	+ 0.185
Distance between the place of residence and the place of education in 1000 kilometres	+ 0.227	− 0.145	+ 5.479	+ 0.468	− 3.810
In employment	+ 1.063 *	− 0.317	− 0.109	+ 1.109 *	+ 0.282
Change in employment	+ 0.069	+ 0.181	+ 0.017	+ 0.167 *	+ 0.120
Distance between the place of residence and the place of employment in 1000 kilometres	− 0.493 *	− 0.217	+ 1.222 *	− 8.174 *	− 1.364 *
Monthly income in 1000 CHF	+ 0.511 *	− 0.345 *	+ 0.030	− 0.111	+ 0.228 *
Monthly income in 1000 CHF squared	− 0.024 *	+ 0.014 *	− 0.000	+ 0.011	− 0.012 *
Car availability: Always			− 2.243 *	− 2.370 *	− 1.720 *
Car availability: Partially			− 0.927 *	− 0.120	− 1.039 *
National ticket ownership	− 3.799 *	+ 0.486			
Regional ticket ownership	− 3.076 *	+ 1.220 *			+ 1.080 *
Half-fare discount ticket ownership	− 2.173 *	+ 0.096		+ 0.878 *	
Moving out of parents' house	+ 0.408	+ 0.062	+ 0.061	+ 0.020	+ 0.118
Change in residence	− 0.015	− 0.054	+ 0.212	− 0.195 *	− 0.007
Birth of a person in the household	+ 0.340	+ 0.119	− 0.567	− 0.073	− 0.269
Number of persons in the household	− 0.286 *	+ 0.150	− 0.163	− 0.131	− 0.009
Number of rooms in the accommodation	+ 0.078	+ 0.038	− 0.050	+ 0.065	+ 0.006
Degree of urbanisation: Urban (referential category)					
Urban to rural	− 0.115	+ 0.479 *	− 0.436	− 0.572 *	+ 0.152
Rural	+ 0.933 *	− 0.379	− 1.863 *	− 0.989 *	+ 0.611
Place of residence abroad	+ 0.573	− 0.395	− 2.705 *	− 2.104 *	− 3.002 *
Purchasing power index in the residential region	− 0.003	+ 0.028 *	+ 0.116 *	+ 0.028 *	+ 0.025 *
Constant	− 11.866 *	− 6.331 *	− 15.689 *	− 3.048 *	− 10.609 *
Individual-specific random parameter	+ 6.999 *	− 5.288 *	+ 7.962 *	− 5.339 *	− 5.550 *

Table 54 is continued ...

Table 54 continued ...

Explanatory variable	Car availability: Always	Car availability: Partially	National ticket ownership	Regional ticket ownership	Half-fare discount ticket ownership
Number of persons	1043	1043	1043	1043	1043
Number of observations	28808	28808	28808	28808	28808
ρ^2 (adjusted)	0.759	0.744	0.859	0.728	0.582

* Level of significance ≤ 0.10

Discrete choice modelling for the ownership of mobility tools in groups

Table 55 shows the mobility tool ownership in six different groups, which cover all possible combinations, by the ARE type of the residential municipality. The largest group with about one third are mere car owners, followed by persons having a car and public transport season tickets at their disposal. Compared to the year 2005, the shares of respondents without any available mobility tools and with only a car are noticeably higher, when considering the 20 year period from 1985 to 2004. The various municipality types differ significantly from one another. For persons living abroad, the share with no mobility tools is with over 30% considerably higher than for Switzerland, where the lowest shares are found in the middle and ancillary centres as well as in the agglomeration municipalities. Mere public transport season tickets owners tend to live in the main centres, while non-ownership is more widespread abroad, in rural areas and in agglomeration municipalities. The share of the group with a car, but no season tickets lies noticeably above the average in other countries, whereas in Switzerland persons tend to own a car and season tickets more frequently simultaneously.

Table 55   Mobility tool ownership in groups by ARE classification (1985-2004)

Spatial and transport system defined classification	No Car + No Tickets	No Car + HF T	No Car + Nat T / Reg T	Car + No Tickets	Car + HF T	Car + Nat T / Reg T
Main centres	7.0%	10.6%	21.8%	25.4%	17.1%	18.1%
Middle and ancillary centres with railway access	3.8%	7.6%	8.2%	36.1%	28.9%	15.4%
Middle and ancillary centres without railway access	5.3%	8.3%	6.4%	41.8%	23.9%	14.3%
Agglomeration municipalities	4.2%	5.2%	6.1%	43.8%	24.8%	15.9%
Rural areas	7.6%	4.2%	5.1%	49.0%	23.1%	10.9%
Abroad	30.9%	4.8%	4.7%	52.9%	1.4%	5.4%
Overall	10.4%	7.4%	11.4%	37.5%	19.4%	13.9%

Table 56 illustrate the changes in mobility tool ownership taking place during the period from 1985 to 2004. In total 1695 transitions between the different groups are observed for the persons aged 18 years and older. In about one third of all cases respondents acquire a car, whereas only 10% are related to the abandonment of a car. To a slightly lesser extent and in a more balanced way this also applies to the various public transport season tickets. Overall, the highest shares arise for the transitions to the alternatives including cars, with about 25% in each case.

Table 56   Changes in mobility tool ownership in groups (1985-2004)

Previous ownership group / Ownership group	No Car + No Tickets	No Car + HF T	No Car + Nat T / Reg T	Car + No Tickets	Car + HF T	Car + Nat T / Reg T	Overall
No Car + No Tickets	0.0%	0.9%	1.6%	2.2%	0.1%	0.4%	5.1%
No Car + HF T	3.1%	0.0%	3.0%	0.5%	1.4%	0.4%	8.3%
No Car + Nat T / Reg T	3.1%	4.5%	0.0%	1.5%	0.7%	2.7%	12.4%
Car + No Tickets	10.7%	0.8%	2.2%	0.0%	4.9%	5.9%	24.5%
Car + HF T	0.8%	5.0%	0.9%	11.8%	0.0%	6.5%	25.0%
Car + Nat T / Reg T	0.5%	0.5%	10.6%	7.1%	5.9%	0.0%	24.6%
Overall	18.3%	11.7%	18.3%	23.0%	12.9%	15.8%	100.0%

In Table 57 the results of a corresponding multinomial logit model for the ownership of mobility tools in the 20 year period from 1985 to 2004 are shown. No individual-specific random parameter is incorporated, since the consideration of the panel effect is not feasible for the nested and cross-nested models, as their estimation runs into numerical problems. So, a comparison of the MNL, NL and CNL models remains possible. However, in Appendix B 2 (Table B 2-3, Table B 2-4) the two MNL models, one not taking the panel effect into account and one taking it into account, are represented. Again, both the estimated individual-specific random parameter is significant and ρ^2 is considerably higher in the second model.

Table 57   Multinomial logit model for mobility tool ownership in groups (1985-2004)

Explanatory variable	No Car + No Tickets	No Car + HF T	No Car + Nat T / Reg T	Car + No Tickets	Car + HF T	Car + Nat T / Reg T
Age in years	– 0.038 *	– 0.028 *	– 0.163 *		– 0.015 *	– 0.090 *
Age in years squared	+ 0.001 *	+ 0.001 *	+ 0.002 *		+ 0.000 *	+ 0.001 *
Gender: Male	– 0.999 *	– 0.939 *	– 1.029 *		– 0.517 *	– 0.552 *
Nationality: Swiss national	– 0.691 *	+ 0.544 *	+ 0.183 *		+ 1.249 *	+ 0.818 *
College or university degree	– 0.095	+ 0.549 *	+ 0.700 *		+ 0.994 *	+ 0.956 *
In education	+ 0.225	+ 0.742 *	+ 0.866 *		+ 0.481 *	+ 1.036 *
Change in education	+ 0.205	+ 0.378 *	+ 0.320 *		+ 0.214 *	+ 0.236 *
Distance between the place of residence and the place of education in 1000 kilometres	+ 1.072	– 4.769	+ 1.985 *		+ 2.253 *	+ 2.355 *
In employment	– 0.887 *	– 0.195 *	– 0.088		+ 0.226 *	+ 0.371 *
Change in employment	– 0.233	– 0.047	+ 0.105		+ 0.105	+ 0.257 *
Distance between the place of residence and the place of employment in 1000 kilometres	+ 0.015	+ 0.012	– 1.395		– 0.144 *	– 0.261 *
Monthly income in 1000 CHF	– 0.088	– 0.075 *	– 0.069 *		+ 0.022	+ 0.058 *
Monthly income in 1000 CHF squared	– 0.037 *	– 0.006 *	+ 0.003 *		+ 0.003 *	– 0.001
Moving out of parents' house				+ 0.146		
Change in residence				– 0.040		
Birth of a person in the household				+ 0.236 *		
Number of persons in the household				– 0.020		
Number of rooms in the accommodation				+ 0.071 *		
Degree of urbanisation: Urban (referential category) Urban to rural Rural				+ 0.728 * + 0.920 *		
Place of residence abroad				+ 1.906 *		
Purchasing power index in the residential region				– 0.014 *		
Constant	– 0.318	– 1.954 *	+ 1.514 *		– 2.905 *	– 0.962 *
Number of observations						28808
ρ^2 (adjusted)						0.214

*   Level of significance ≤ 0.10

Overall, age has a positive effect with reference to the group of mere car owners, especially for the older respondents. Only the alternatives including national and regional season tickets show a slight decrease of the utility for younger persons. Gender has a negative influence. This means that men tend to merely own a car more frequently than women. Respondents without any mobility tools are more likely to be foreign national as well as not to hold a college or university degree. The opposite tendency is visible for the other groups which own mobility tools. For these groups, being in education, a change in education and the corresponding distance show a positive influence. Persons with no mobility tools or with only a half-fare discount ticket are less likely to be employed, whereas employment and a change in employment increases the probability of having a car and public transport season tickets simultaneously at disposal. At the same time, the distance between the place of residence and the place of employment has a negative effect for these last two groups. With increasing income the propensity to not have a mobility tool or to not have a car decreases. For the respondents with an available car and public transport season tickets, the mean elasticity concerning the monthly income in 1000 CHF is positive and amounts to approximately 0.2. Mere car owners tend to live in more rural areas as well as not in Switzerland. Thereby, the index of purchasing power in the residential region has a negative influence on car ownership.

Table 58 presents the parameters of a nested logit model with two nests regarding the ownership and non-ownership of a car, i.e., the same structure as shown in Figure 13 is used. For the alternatives including the national and regional tickets, age has a negative effect until the age of about 50 years. Afterwards, the utility increases with increasing age. With regards to the other alternatives, age leads to a higher propensity to choose one of these. Overall, men are considerably more likely to be mere car owners than women. With the exception of the persons with no mobility tools, being a Swiss national as well as holding a college or university degree has a positive influence. This also applies to education and employment as well as to changes occurring in education and employment. The distance between the places of residence and education increases the probability of mobility tool ownership, except for the half-fare discount tickets, in reference to only car owners. Concerning the place of employment, the distance has in general a negative effect. A higher income enhances the simultaneous availability of cars and public transport season tickets. The birth of a person in the household as well as the household size and accommodation size increase the ownership of a car. These owners are primarily found in more rural areas. Overall, the nested logit model is relatively similar to the corresponding multinomial logit model shown in Table 57. At the same time, the measure for the goodness of fit only shows a minor rise. The scale parameters estimated in the NL model are both significant. They indicate that the correlations in the nest with the alternatives including a car are slightly smaller than in the other nest.

Table 58 Nested logit model for mobility tool ownership in groups with two nests for car and no car (1985-2004)

Explanatory variable	No Car + No Tickets	No Car + HF T	No Car + Nat T / Reg T	Car + No Tickets	Car + HF T	Car + Nat T / Reg T
Age in years	− 0.056	+ 0.107 *	− 0.420 *		− 0.081 *	− 0.297 *
Age in years squared	+ 0.001 *	− 0.000	+ 0.005 *		+ 0.002 *	+ 0.003 *
Gender: Male	− 1.645 *	− 1.866 *	− 1.532 *		− 1.309 *	− 2.138 *
Nationality: Swiss national	− 1.264 *	+ 2.656 *	+ 1.530 *		+ 4.678 *	+ 2.943 *
College or university degree	− 0.361	+ 2.162 *	+ 2.308 *		+ 3.663 *	+ 3.206 *
In education	− 0.746	+ 2.211 *	+ 2.224 *		+ 1.203 *	+ 3.719 *
Change in education	+ 0.317	+ 1.143 *	+ 0.895 *		+ 0.953 *	+ 1.263 *
Distance between the place of residence and the place of education in 1000 kilometres	+ 5.341	− 27.082 *	+ 7.436 *		+ 8.128 *	+ 8.036 *
In employment	− 1.787 *	+ 0.452 *	+ 0.566 *		+ 0.709 *	+ 1.332 *
Change in employment	− 0.810 *	+ 0.286	+ 0.564 *		+ 0.348	+ 1.102 *
Distance between the place of residence and the place of employment in 1000 kilometres	+ 0.292	− 0.089	− 3.598 *		− 0.186	− 1.129 *
Monthly income in 1000 CHF	− 0.551 *	+ 0.199 *	+ 0.049		+ 0.008	+ 0.256 *
Monthly income in 1000 CHF squared	− 0.025	− 0.028 *	+ 0.012 *		+ 0.014 *	− 0.005
Moving out of parents' house				+ 0.627 *		
Change in residence				+ 0.010		
Birth of a person in the household				+ 0.815 *		
Number of persons in the household				+ 0.009		
Number of rooms in the accommodation				+ 0.199 *		
Degree of urbanisation: Urban (referential category) Urban to rural Rural				+ 1.961 * + 2.214 *		
Place of residence abroad				+ 3.148 *		
Purchasing power index in the residential region				− 0.007 *		
Constant	+ 2.463 *	− 7.216 *	+ 6.611 *		− 6.798 *	− 0.583

Model parameters for the two nests:	
Nest: Car	0.247 *
Nest: No Car	0.305 *
Number of observations	28808
ρ^2 (adjusted)	0.217

* Level of significance ≤ 0.10

In Table 59 a cross-nested logit model is represented, differentiating between four nests for car, national and regional tickets, half-fare discount tickets as well as no mobility tools. The corresponding model structure is shown in Figure 14. This model also very much resembles the multinomial logit model and the nested logit model, especially concerning the way in which the explanatory variables influence the choices of the alternatives. Therefore, the corresponding results of the CNL model are not commented on again. At the same time, the majority of the estimated model parameters describing the cross-nested structure are significant. The highest correlations occur among the alternatives in the nest regarding the half-fare discount tickets. The parameters indicating the degree at which an alternative belongs to a certain nest is, concurrent with the expectations, always largest for the alternatives that merely include the mobility tool defining the nest and, thus, do not belong to any other nest. The group owning a car and a half-fare discount ticket is primarily part of the nest for cars and to a lesser extent part of the nest for the half-fare discount tickets. In contrast to that, the respondents having a car as well as national and regional tickets at their disposal predominantly belong to the nest for national and regional tickets and not at all to the nest for cars. Nevertheless, the adjusted ρ^2 is considerably lower in the CNL model than in the MNL and NL models.

Table 59   Cross-nested logit model for mobility tool ownership in groups with four nests for car, national and regional tickets, half-fare discount tickets and no mobility tools (1985-2004)

Explanatory variable	No Car + No Tickets	No Car + HF T	No Car + Nat T / Reg T	Car + No Tickets	Car + HF T	Car + Nat T / Reg T
Age in years	+ 0.085 *	− 0.009	− 0.135 *		− 0.035 *	− 0.067 *
Age in years squared	− 0.001 *	+ 0.000 *	+ 0.001 *		+ 0.000 *	+ 0.000 *
Gender: Male	− 1.044 *	− 1.136 *	− 1.693 *		− 0.718 *	− 0.489 *
Nationality: Swiss national	− 0.735 *	+ 0.824 *	+ 0.453 *		+ 1.766 *	+ 1.009 *
College or university degree	− 0.295 *	+ 0.841 *	+ 0.906 *		+ 1.341 *	+ 1.153 *
In education	+ 0.250 *	+ 0.340 *	+ 1.352 *		+ 0.162	+ 0.662 *
Change in education	− 0.003	+ 0.190	+ 0.383 *		− 0.015	+ 0.068
Distance between the place of residence and the place of education in 1000 kilometres	+ 2.740	+ 5.090 *	+ 3.334		− 0.879	+ 4.372
In employment	− 1.378 *	− 0.187 *	− 0.316 *		+ 0.094	+ 0.642 *
Change in employment	− 0.218	− 0.007	+ 0.280 *		+ 0.043	+ 0.268 *
Distance between the place of residence and the place of employment in 1000 kilometres	− 0.474	− 0.746	− 1.025		− 0.668	− 25.292 *

Table 59 is continued ...

Table 59 continued ...

Explanatory variable	No Car + No Tickets	No Car + HF T	No Car + Nat T / Reg T	Car + No Tickets	Car + HF T	Car + Nat T / Reg T
Monthly income in 1000 CHF	− 0.322 *	− 0.232 *	− 0.105 *		+ 0.013	+ 0.117 *
Monthly income in 1000 CHF squared	− 0.017	+ 0.009 *	+ 0.008 *		+ 0.000	− 0.016 *
Moving out of parents' house				+ 0.043		
Change in residence				− 0.032		
Birth of a person in the household				+ 0.276 *		
Number of persons in the household				− 0.101 *		
Number of rooms in the accommodation				+ 0.153 *		
Degree of urbanisation: Urban (referential category)						
Urban to rural				+ 1.002 *		
Rural				+ 1.025 *		
Place of residence abroad				+ 2.084 *		
Purchasing power index in the residential region				− 0.011 *		
Constant	− 0.389	− 0.387 *	+ 1.180 *		− 1.268 *	− 0.289

Model parameters for the four nests as well as for the six groups:	
Nest: Car	0.963
Car + No Tickets	1.324 *
Car + HF T	0.617 *
Car + Nat T / Reg T	0.000 *
Nest: National and regional tickets	0.578 *
No Car + Nat T / Reg T	1.476 *
Car + Nat T / Reg T	0.971 *
Nest: Half-fare discount tickets	2.057 *
No Car + HF T	1.082 *
Car + HF T	0.289 *
Nest: No mobility tools	1.000
No Car + No Tickets	0.867 *
Number of observations	28808
ρ^2 (adjusted)	0.177

* Level of significance ≤ 0.10

Table 60 shows the results of a multivariate probit model for the mobility tool ownership in the six different groups. Concerning the respondents without any mobility tools at their disposal, age has a negative effect until a minimum is reached at the age of 40 years. Afterwards, the utility increases with increasing age. This also applies to the other alternatives, with the group only owning a half-fare discount ticket reaching this minimum earlier during the life course, while the other groups reach it later. Gender has a negative influence. This means that men more frequently tend to merely own a car than women. Persons having both a car and public transport season tickets available are more likely to be Swiss and to have a college or university degree with reference to the group of mere car owners. In contrast, respondents with no mobility tools tend to be foreign nationals, not holding a higher educational degree. In this group education, employment and corresponding changes as well as the income per month show a negative influence. Regarding the other alternatives, education, employment and income do not have such a distinct effect. Employment increases the probability of having a car and public transport season tickets simultaneously at disposal. At the same time, the income tends to be higher in these two groups. Mere car owners are more likely to live in more rural areas as well as abroad. Thereby, the index of purchasing power in the residential region has a negative influence on car ownership. The correlations between the different groups, which are incorporated in the model, are in all cases significant. In this context, the largest values are observed among the alternatives including a car. In comparison to the logit models, the goodness of fit measure ρ^2 improves considerably.

Table 60   Multivariate probit model for mobility tool ownership in groups (1985-2004)

Explanatory variable	No Car + No Tickets	No Car + HF T	No Car + Nat T / Reg T	Car + No Tickets	Car + HF T	Car + Nat T / Reg T
Age in years	− 0.022 *	− 0.056 *	− 0.057 *		− 0.065 *	− 0.075 *
Age in years squared	+ 0.000 *	+ 0.001 *	+ 0.001 *		+ 0.001 *	+ 0.001 *
Gender: Male	− 0.299 *	− 0.298 *	− 0.299 *		− 0.098 *	− 0.099 *
Nationality: Swiss national	− 0.502 *	+ 0.001	− 0.001		+ 0.302 *	+ 0.199 *
College or university degree	− 0.301 *	− 0.002	+ 0.096 *		+ 0.296 *	+ 0.295 *
In education	− 0.301 *	− 0.200 *	+ 0.298 *		− 0.301 *	+ 0.199 *
Change in education	− 0.000	+ 0.000	+ 0.100 *		− 0.099 *	− 0.000
Distance between the place of residence and the place of education in 1000 kilometres	− 1.322 *	− 0.922 *	−1.280 *		− 2.733 *	+ 2.268 *
In employment	− 0.500 *	− 0.096 *	− 0.096 *		+ 0.106 *	+ 0.203 *
Change in employment	− 0.200 *	− 0.100 *	+ 0.001		− 0.098 *	+ 0.100 *
Distance between the place of residence and the place of employment in 1000 kilometres	− 0.074	− 9.430 *	− 0.318		+ 0.028	+ 0.075
Monthly income in 1000 CHF	− 0.103 *	− 0.091 *	+ 0.005		+ 0.009 *	+ 0.098 *
Monthly income in 1000 CHF squared	− 0.004 *	+ 0.004 *	− 0.001 *		+ 0.003 *	− 0.007 *

Table 60 is continued ...

Table 60 continued ...

Explanatory variable	No Car + No Tickets	No Car + HF T	No Car + Nat T / Reg T	Car + No Tickets	Car + HF T	Car + Nat T / Reg T
Moving out of parents' house				− 0.100 *		
Change in residence				− 0.001		
Birth of a person in the household				+ 0.200 *		
Number of persons in the household				+ 0.006 *		
Number of rooms in the accommodation				− 0.010 *		
Degree of urbanisation: Urban (referential category) Urban to rural Rural				+ 0.396 * + 0.498 *		
Population in the residential municipality in 1000 inhabitants				+ 0.001 *		
Population density in the residential municipality in 1000 inhabitants per square kilometre				− 0.076 *		
Place of residence abroad				+ 0.898 *		
Purchasing power index in the residential region				− 0.009 *		
Constant				+ 0.299 *		
Correlation matrix:						
No Car + No Tickets	+ 1.000	− 0.098 *	− 0.091 *	− 0.195 *	− 0.095 *	− 0.093 *
No Car + HF T		+ 1.000	− 0.098 *	− 0.209 *	− 0.199 *	− 0.097 *
No Car + Nat T / Reg T			+ 1.000	− 0.298 *	− 0.194 *	− 0.198 *
Car + No Tickets				+ 1.000	− 0.404 *	− 0.314 *
Car + HF T					+ 1.000	− 0.218 *
Car + Nat T / Reg T						+ 1.000
Number of observations						28808
ρ^2 (adjusted)						0.510

* Level of significance ≤ 0.10

Table 61 describes the various discrete choice models with their initial and final log-likelihood as well as the number of estimated parameters. Again, the different logit models are relatively similar to one another, with the nested logit model being the best model overall. This indicates that structuring the sample by the ownership and non-ownership of cars is most appropriate in comparison to the public transport season tickets, pointing to a stronger commitment towards cars. The log-likelihood ratio G^2 is highest in the multivariate probit model.

Table 61 Description of the various models for mobility tool ownership in groups (1985-2004)

Models for mobility tool ownership in groups	Initial log-likelihood	Final log-likelihood	Number of parameters	ρ^2 (adjusted)
Multinomial logit model	−51617.0	−40503.7	79	0.214
Nested logit model with two nests for car and no car	−51617.0	−40329.7	81	0.217
Cross-nested logit with four nests for car, national and regional tickets, half-fare discount tickets and no mobility tools	−51617.0	−42383.3	91	0.177
Multivariate probit model	119809.1	−58563.5	92	0.510

The various logit models strongly resemble one another with respect to the significant influencing variables as well as with respect to the scale of the estimated parameters. Age, gender, nationality, holding a college or university degree, occupation related variables and income as well as variables describing the residential municipality and region have the most influence. The probit model again differs from the logit models, since the correlations between the different alternatives concerning the mobility tool ownership are incorporated in the probit model.

8.4 Duration modelling for long-term and mid-term mobility

In the following chapter, event history analyses are applied to the retrospective data, on the one hand for the residential mobility and the changing locations of education and employment as well as on the other hand for the ownership of the different mobility tools. In Figure 30 the distribution of the residential, education and employment durations during the 20 year period from 1985 to 2004 is represented, including both uncensored and censored durations. Overall, 4155 residential, 1290 education and 2589 employment durations are observed between 1985 and 2004. On average these durations are 5.0, 3.9 and 4.8 years long with a standard deviation of 4.8, 3.0 and 4.8 years, respectively. Approximately 70% of all the durations are up to five years long.

Figure 31 illustrates the observed durations of car availability and public transport season ticket ownership. For about one third of the durations, cars are always available from 1985 to 2004. In this context, the other duration lengths are relatively evenly distributed. Partial car availability is more often indicated for shorter periods of time with over 50% being less than five years long and over 80% being less than ten years long. The ownership of national and regional tickets shows a left-skewed distribution, where the highest shares occur for durations shorter than five years. To a lesser extent this also applies to the half-fare discount ticket ownership. Comparing the single durations to the durations summarised over the 20 year period, there are only small differences noticeable, primarily concerning the durations from half to three years (Beige, 2006). Overall, the ownership of the different mobility tools is relatively stable over time, especially the availability of cars, whereas the slightly more variable ownership of season tickets during the period from 1985 to 2004 points to a weaker commitment to public transport. This stability in mobility tool ownership over longer periods of time is also found in other studies (Axhausen and Beige, 2003; Bjørner and Leth-Petersen, 2005; Lanzendorf, 2006; Prillwitz et al., 2006; Simma and Axhausen, 2003).

The distribution of mobility tool ownership durations is to a lesser extent left-skewed compared to the durations concerning the places of residence, education and employment. The various groups of observed durations are significantly different from one another. The shortest periods are observed between moves as well as between changes in education and employment, whereas always available cars and half-fare discount tickets stand at the other end of the spectrum. At the same time, the share of censored observations, i.e., where information about the duration is incomplete, as no further transition is made within the surveyed time period, is lowest in the first group.

Figure 30   Distribution of the residential, education and employment durations (1985-2004)

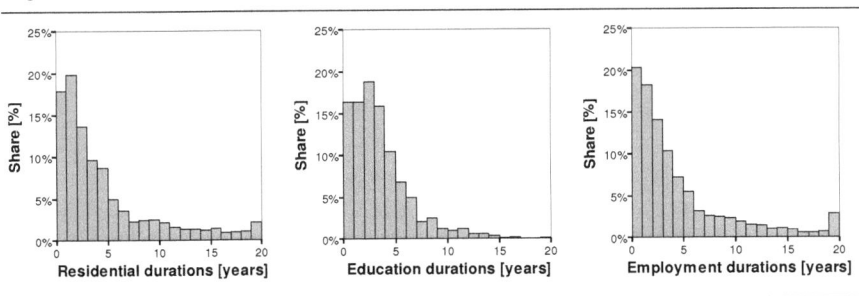

Figure 31 Distribution of the car availability and public transport season ticket ownership durations (1985-2004)

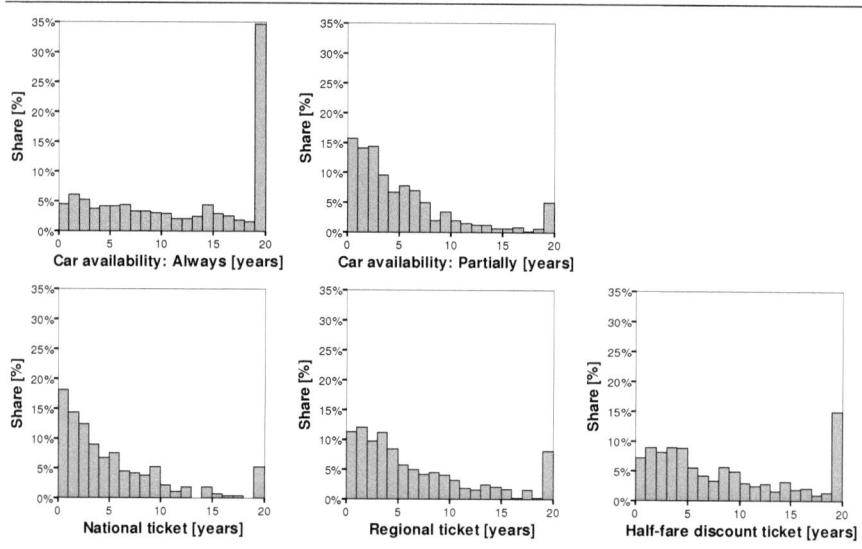

In Figure 32 the hazard rates for the residential, education and employment durations are shown. The hazard rate represents the probability or intensity of events occurring per time unit. In the case of residential mobility, for instance, this is the number of moves per year. For the first three years, the various groups behave very similar with an increase of the hazard to a maximum of about 0.20 changes per year. The probability then slowly decreases, with the exception of education. The relatively strong variations occurring for the longer durations are connected to a relatively small number of observations in this range.

Figure 32   Hazard rates of the duration models for the residential, education and employment durations (1985-2004)

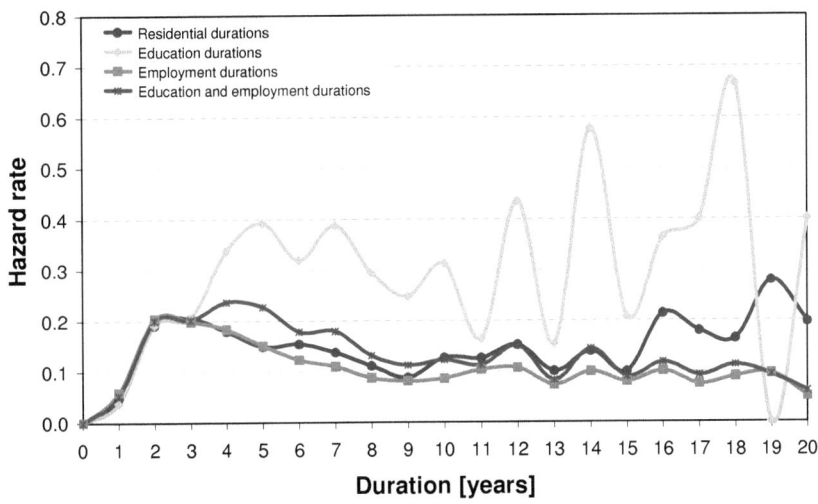

Figure 33 illustrates the hazard rates for the car availability and public transport season ticket ownership durations. The curves for the different mobility tool durations are in comparison to the residential, education and employment durations slightly flatter. This applies especially to the always available cars. At the same time, the hazard rates strongly vary over time, so there are no clear tendencies noticeable.

Figure 33   Hazard rates of the duration models for the car availability and public transport season ticket ownership durations (1985-2004)

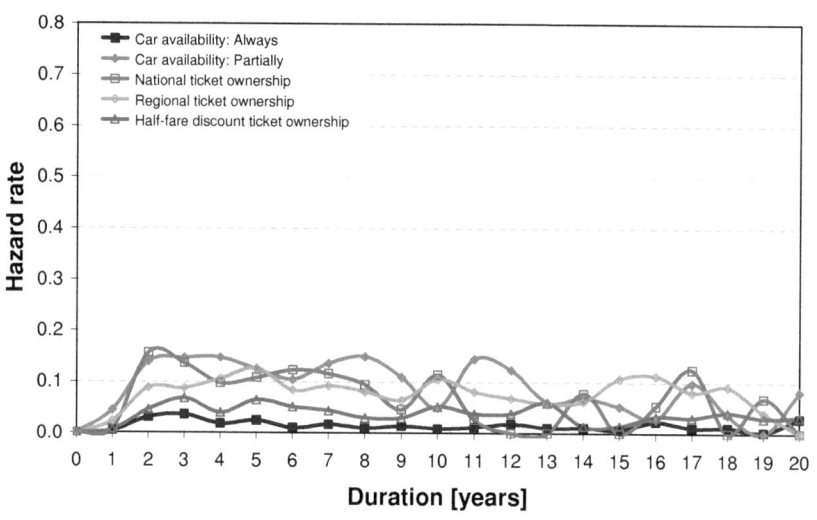

Table 62 represents results of different hazard models for the residential, education and employment durations. All explanatory variables shown are significant at a level of at least 0.10. Their selection is based on a forward stepwise inclusion method using the significance of the change in the log-likelihood as entry or removal criteria. In the table the hazard ratios are shown which are equivalent to the exponential hazard parameters (Allison, 1995). For continuous variables, they indicate the percentage change of the hazard rate, whereas for dichotomous variables they equal the proportion of the two corresponding hazard rates. As a measure of how good the different models are and how well the corresponding durations are predicted with the set of covariates, respectively, generalised R^2's are given at the bottom of the table (Allison, 1995). The generalised R^2 is calculated, as proposed by Cox and Snell, as follows

$$R^2 = 1 - e^{-\frac{2(L(\max) - L(0))}{N}}, \qquad (8\text{-}1)$$

where $L(0)$ and $L(max)$ represent the initial and the final log-likelihoods, respectively, and N is the sample size. In the estimated models shown in the table the durations are relatively well predictable by the given explanatory variables.

The variable for the occurrence of left censoring has a positive influence on the duration, concurrent with the expectations. With increasing age persons tend to move less and to stay longer at a place of residence. This result is consistent with most of the literature (Clark and Onaka, 1983; Courgeau, 1985; Vandersmissen *et al.*, 2005). For example, Hollingworth and Miller (1996) found that age is the variable with the most distinct effect on residential mobility. The gender of the respondents has no influence. Respondents with a college or university degree are more likely to change their place of residence. Changes in education and employment during the observed period lead to a lower probability of moving, contradicting findings of Hollingworth and Miller (1996), Rouwendal and van der Vlist (2005), Van der Waerden *et al.* (2003), Verhoeven *et al.* (2005), and others. The share in education affects the duration in employment positively, and vice versa. The distances between the places of residence, education and employment increase the various hazards of changes occurring considerably. So, persons seem to compensate the corresponding distances by shorter stays. The ownership of the different mobility tools, especially the national and regional tickets, shows positive hazard parameters, indicating that more mobile persons are, at the same time, also more spatially mobile. Simultaneous changes of the places of residence, education and employment strongly increase the hazard of changes occurring, thereby confirming results in the literature (Lelièvre and Bonvalet, 1994). Moving out of the parents' house leads to a longer stay at the following place of residence. The number of moves has a positive effect on the durations of education and employment. The number of births in the household leads to longer durations between moves. With each birth the hazard rate decreases by about 30.9%. The household size and the accommodation size affect the various durations positively as well. This also applies to the education and employment durations. In the case that the place of residence is abroad, the observed durations tend to be considerably shorter. The index of purchasing power in the residential region has a hazard ratio that is smaller than one, thereby indicating an increase in the probability of changes occurring.

Table 62 Hazard ratios of the duration models for the residential, education and employment durations (1985-2004)

Explanatory variable *(Average values for the observed durations)*	Residential durations	Education durations	Employment durations	Education and employment durations
Left censoring of the duration	0.368	0.533	0.425	0.482
Age in years			0.871	
Age in years squared	0.999		1.001	0.999
Age in years natural logarithm	1.467			
Age in years * Gender: Male		0.993		
Nationality: Swiss national			1.311	1.209
College or university degree	1.189		1.127	
Share in education during the period			0.742	
Duration in education at the beginning of the period in years	1.036			
Changes in education during the period	0.706		0.874	0.893
Distance between the place of residence and the place of education in 1000 kilometres	1.877	1.224		1.199
Share in employment during the period		0.711		1.361
Changes in employment during the period	0.772			
Distance between the place of residence and the place of employment in 1000 kilometres		118.324	1.218	1.254
Monthly income squared		1.004		
Monthly income natural logarithm		1.228		
Car availability: Always				1.260
Car availability: Partially				1.328
National ticket ownership	1.470	1.610		1.603
Regional ticket ownership	1.181	1.302		1.215
Half-fare discount ticket ownership				1.156
Simultaneous change of the place of residence and the places of education or employment	1.709	1.819	1.543	1.688
Moving out of parents' house	0.834		0.816	0.569
Duration in residence at the beginning of the period in years			0.991	
Changes in residence during the period		0.618	0.564	
Number of births in the household	0.691	0.670	0.632	0.595
Number of persons in the household	0.930		0.918	0.950
Number of rooms in the accommodation	0.944	0.957		0.955
Place of residence abroad	1.219		1.653	1.581
Purchasing power index in the residential region	0.970	0.985	0.975	0.976

Table 62 is continued ...

Table 62 continued ...

Explanatory variable *(Average values for the observed durations)*	Residential durations	Education durations	Employment durations	Education and employment durations
Number of observations	3511	1078	2291	3369
Number of censored observations	1039	92	788	880
R^2 (generalised)	0.378	0.185	0.416	0.351

In Table 63 the hazard ratios of the different models for the mobility tool ownership durations are shown. The generalised R^2 are to some extent larger than in the models for the residential, education and employment durations. Again, left censoring has a strongly negative effect on the various hazards. The mobility tool ownership durations are positively influenced by the age of the respondents, especially this is true for cars as well as for national and regional tickets, confirming the general hypothesis that over the life course younger adults tend to be more open to change than the elderly. Men are more likely to hold a national ticket over longer periods of time than women. Changes of the places of residence, education and employment during the observed period reduce the hazard significantly. Longer distances from the place of residence to the place of education lead to shorter durations of public transport season ticket ownership. Persons show with increasing income a higher stability for always available cars and a lower stability for partially available cars. Fuel prices affect the durations considerably in a positive way. The ownership of a half-fare discount ticket increases the hazard for always available cars. When a simultaneous change of residence, education and employment occurs, the ownership durations tend to be much shorter. The same applies to the moves out of the parents' house and the effect on car availability. The number of births in the household reduces the probability of variations in mobility tool ownership. Concerning the degree of urbanisation, the durations for which persons own a half-fare discount ticket are significantly shorter in the urban areas than in the rural areas. With an increasing index of purchasing power the hazard rate for national tickets decreases.

Table 63   Hazard ratios of the duration models for the car availability and public transport season ticket ownership durations (1985-2004)

Explanatory variable *(Average values for the observed durations)*	Car availability: Always	Car availability: Partially	National ticket ownership	Regional ticket ownership	Half-fare discount ticket ownership
Left censoring of the duration	0.201	0.253	0.130	0.364	0.208
Age in years	0.912	0.926			
Age in years squared			0.998	0.999	
Age in years natural logarithm					0.165
Gender: Male			0.674		
Share in education during the period				2.130	
Duration in education at the beginning of the period in years					1.055
Changes in education during the period	0.720	0.721	0.508	0.665	0.732
Distance between the place of residence and the place of education in 1000 kilometres			1.468	48.951	1.434
Share in employment during the period				2.194	
Duration in employment at the beginning of the period in years	1.056				
Changes in employment during the period		0.814	0.751	0.849	0.796
Monthly income squared		1.008	1.009		
Monthly income natural logarithm	0.517				
Fuel price in 0.01 CHF per litre (lead free 95)	0.940	0.966		0.966	0.942
Half-fare discount ticket ownership	1.585				
Simultaneous change of the place of residence and the places of education or employment	4.008		2.482	1.598	1.965
Moving out of parents' house	0.416	0.623			
Changes in residence during the period	0.535	0.782	0.641	0.652	0.758
Number of births in the household	0.530	0.701		0.652	0.563
Number of rooms in the accommodation	0.854			0.903	

Degree of urbanisation:
    Urban (referential category)
    Urban to rural
    Rural
					0.692
					0.607

Purchasing power index in the residential region			0.963		
Number of observations	775	406	241	515	748
Number of censored observations	640	161	114	217	478
R^2 (generalised)	0.376	0.510	0.552	0.471	0.457

In order to compare the different types of durations, competing risks models for the residential, education and employment durations on the one hand as well as for the car availability and public transport season ticket ownership durations on the other hand are estimated. For each type of duration, models are estimated treating the others in this context as right censored (Allison, 1995; Box-Steffensmeier and Jones, 2004). In Figure 34 the hazard rates for the residential, education and employment durations are shown. The probability for moving rises for the first three years to a maximum of about 0.10 moves per year. It then slowly decreases and for durations that are ten years and longer increases again. In the range between 15 and 20 years the hazard rate for the residential mobility varies relatively strongly. The education and employment durations show a very similar trend though at a lower level. Overall, these hazard rates are considerably lower than the ones of the basic duration models which are presented in Figure 32.

Figure 34   Hazard rates of the competing risks duration models for the residential, education and employment durations (1985-2004)

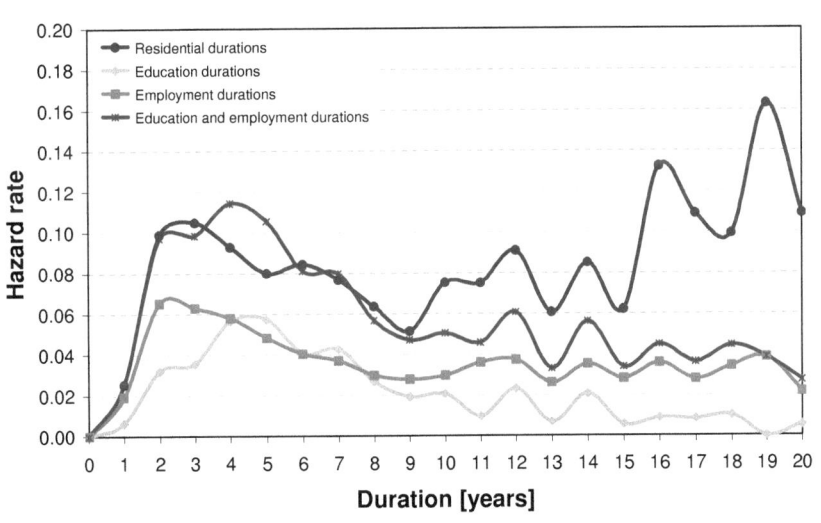

Figure 35 shows the hazard rates for the car availability and public transport season ticket ownership durations in the competing risks situation. The curves for the different mobility tool durations are in comparison to the residential, education and employment durations very flat with the hazard rate not rising above a 0.02 level. There are no clear tendencies visible. The hazard rates strongly vary over time.

Figure 35 Hazard rates of the competing risks duration models for the car availability and public transport season ticket ownership durations (1985-2004)

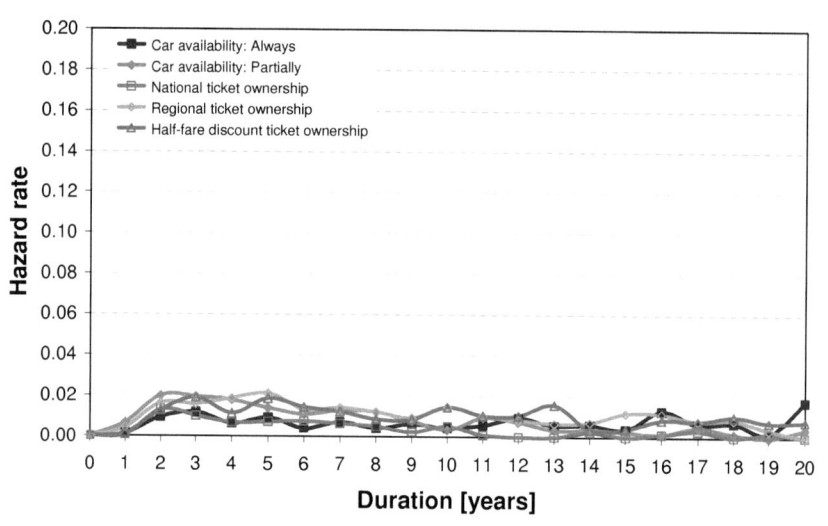

Table 64 presents the results of the different competing risks models for the residential, education and employment durations, grouping the observations for these three types together. The latent duration time approach is applied, which means that for each specific type a model is estimated, treating the other types as censored (Box-Steffensmeier and Jones, 2004). In the table the hazard ratios are given. The variable indicating that the duration is left censored has a strong positive influence. With increasing age the hazard of changes occurring in residence, education and employment decreases. In this context, men are by about 9.5% less likely to move than women. Respondents holding a college or university degree tend to move more frequently. Persons in education and employment are at a lower risk of changing the place of residence. At the same time, education leads to a considerably lower hazard in employment, and vice versa. A longer duration in education at the beginning of the period shortens the various durations, whereas a longer duration in employment prolongs the corresponding duration in employment. Changes in education and employment during the observed period have a negative influence on the propensity to move. The education and employment durations show opposite effects concerning the number of changes. Respondents with many changes in education are less likely to change education, but more likely to change employment. For the changes in employment, it is the other way around. The distances between the places of residence, education and employment increase the probability of changes in education and employment, respectively. The residential durations are negatively affected by the monthly income. The ownership of the different mobility tools leads to higher hazards regarding spatial mobility. Simultaneous changes of the places of residence, education and employment strongly increase the probability of variations. The duration a person already lives in a place has a positive influence on the residential duration, which is primarily connected to the cases where left censoring occurs, and a negative influence on the durations in education and employment. The number of births as well as the size of the household and the accommodation reduces the various risks. Abroad the durations in residence, education and employment tend to be by over 40% shorter than in Switzerland. The index of purchasing power in the residential region has a hazard ratio that is smaller than one.

Table 64   Hazard ratios of the competing risks duration models for the residential, education and employment durations (1985-2004)

Explanatory variable *(Average values for the observed durations)*	Residential durations	Education durations	Employment durations	Education and employment durations
Left censoring of the duration	0.598	0.428	0.316	0.320
Age in years	1.043	1.125		1.025
Age in years squared	0.999	0.997	0.999	0.999
Gender: Male	0.905			
Nationality: Swiss national		1.308		1.240
College or university degree	1.211			
Share in education during the period	0.605		0.087	
Duration in education at the beginning of the period in years	1.072	1.070	1.074	
Changes in education during the period	0.891	0.013	1.130	0.556
Distance between the place of residence and the place of education in 1000 kilometres		1.241		1.205
Share in employment during the period	0.660	0.116		
Duration in employment at the beginning of the period in years	1.025	1.055	0.974	0.977
Changes in employment during the period	0.832	1.196	0.032	0.340
Distance between the place of residence and the place of employment in 1000 kilometres			1.289	1.328
Monthly income natural logarithm	1.182	0.839		
Car availability: Always			1.526	1.241
Car availability: Partially			1.855	1.460
National ticket ownership	1.420	1.750	1.928	1.715
Regional ticket ownership		1.242	1.455	1.331
Half-fare discount ticket ownership		1.375	1.254	1.264

Table 64 is continued ...

Table 64 continued ...

Explanatory variable (Average values for the observed durations)	Residential durations	Education durations	Employment durations	Education and employment durations
Simultaneous change of the place of residence and the places of education or employment	2.125	2.213	2.030	1.682
Moving out of parents' house		0.640	0.582	0.592
Duration in residence at the beginning of the period in years	0.943	1.046	1.032	1.036
Changes in residence during the period	0.001		0.832	
Number of births in the household	0.735	0.381	0.526	0.479
Number of persons in the household	0.952		0.865	0.928
Number of rooms in the accommodation	0.882			
Place of residence abroad	1.495		1.486	1.428
Purchasing power index in the residential region	0.975	0.975	0.966	0.968
Number of observations	6880	6880	6880	6880
Number of censored observations	4408	5894	5377	4391
R^2 (generalised)	0.402	0.346	0.313	0.325

In Table 65 the hazard ratios of the different competing risks models for the mobility tool ownership durations are shown, when all mobility tools are grouped together. Again, left censoring has a strongly positive effect on the durations. With increasing age the respondents tend to own mobility tools longer. This is especially true for older persons. The share spent in education during the observed periods has a negative influence on the hazard for always available cars. Changes in residence, education and employment decrease the probability of variations in the ownership of mobility tools, whereas in the case that these changes occur simultaneously, the probability is considerably increased. Higher fuel prices lead to reduced hazards. Contrary to the expectations, this also applies to the ownership of the different mobility tools in the competing risks situation. This means that the longer respondents hold one mobility tool the less likely they are to change the ownership of any other mobility tool, since the corresponding hazard ratios are distinctly below one in these cases, especially for car availability. The number of births in the household affects the various durations positively. National tickets are owned for shorter periods of time generally by Swiss nationals, persons with a college or university degree and persons living in non-urban areas.

Table 65 Hazard ratios of the competing risks duration models for the car availability and public transport season ticket ownership durations (1985-2004)

Explanatory variable *(Average values for the observed durations)*	Car availability: Always	Car availability: Partially	National ticket ownership	Regional ticket ownership	Half-fare discount ticket ownership
Left censoring of the duration	0.388	0.250	0.114	0.309	0.233
Age in years	0.919	1.175	1.298	0.932	0.966
Age in years squared		0.996	0.995		
Nationality: Swiss national			2.598		
College or university degree			1.795		
Share in education during the period	0.332			1.775	
Duration in education at the beginning of the period in years	1.064	1.139			
Changes in education during the period		0.741	0.708	0.776	
Distance between the place of residence and the place of education in 1000 kilometres			1.614		
Share in employment during the period				1.985	
Duration in employment at the beginning of the period in years		1.086			
Changes in employment during the period		0.832	0.619	0.863	0.809
Distance between the place of residence and the place of employment in 1000 kilometres			5.600		
Monthly income natural logarithm		0.738			
Fuel price in 0.01 CHF per litre (lead free 95)	0.954	0.956		0.958	0.950
Car availability: Always			0.173	0.295	0.286
Car availability: Partially			0.328	0.296	0.261
National ticket ownership	0.283	0.546			
Regional ticket ownership	0.276	0.441			0.303
Half-fare discount ticket ownership	0.474	0.565		0.664	
Simultaneous change of the place of residence and the places of education or employment	2.284	2.677	1.929	1.918	1.857
Moving out of parents' house		0.564			
Duration in residence at the beginning of the period in years	0.964			0.978	
Changes in residence during the period	0.537	0.618	0.722	0.584	0.679
Number of births in the household	0.639	0.755	0.336	0.563	0.533
Number of persons in the household	0.802				
Number of rooms in the accommodation	0.844			0.913	

Table 65 is continued ...

Table 65 continued ...

Explanatory variable *(Average values for the observed durations)*	Car availability: Always	Car availability: Partially	National ticket ownership	Regional ticket ownership	Half-fare discount ticket ownership
Degree of urbanisation:					
Urban (referential category)					
Urban to rural			0.646		
Rural			0.538		
Number of observations	2685	2685	2685	2685	2685
Number of censored observations	2550	2440	2558	2387	2415
R^2 (generalised)	0.100	0.203	0.116	0.235	0.172

Another possibility to compare the different durations is the estimation of a combined model for all the durations. In Table 66 the corresponding results of various models are represented. The first one, in contrast to the second one, does not include a specification for the type of duration, treating all durations alike. The last model employs the so-called fixed-effects partial likelihood method (FEPL), which allows for correcting for unobserved heterogeneity by introducing a disturbance term (Allison, 1995; Box-Steffensmeier and Jones, 2004). In this context, the duration type is included as stratification variable. Therefore, each stratum has its own baseline hazard rate, while the hazard parameters are restricted to be the same across strata. One disadvantage is that there are no estimates obtained for the effect of the stratifying variable. At the same time, it is not legitimate to compare the log-likelihoods for the models with and without a stratifying variable (Allison, 1995). All the models shown in Table 66 have relatively high generalised R^2's. Overall, the estimated hazard parameters in the three models are very similar with only rather small deviations. A few differences concerning the significance and non-significance of variables occur, for instance, the degree of urbanisation is only in the last model of any importance. The estimated parameters for the variable describing the type reflect the length of the various durations. The longest durations are observed for always available cars, followed by half-fare discount tickets. Hence, the corresponding hazard ratios are the smallest ones. Left censoring has, as one expects, a positive effect on the durations. With increasing age the hazard decreases. Until the age of about 40 years men are at a lower risk than women. Afterwards, this tendency is reversed. Swiss nationals show a higher hazard as well as persons holding a college or university degree. The durations in education and employment at the beginning of the period increase the probability of variations, whereas the number of changes in education and employment during the period reduce it. The distances between the place of residence and the places of education and employment have a negative influence on the durations. With increasing monthly income the hazard rises as well until it reaches a maximum between 2700 CHF and

3000 CHF. Then, the hazard rate slowly declines. The ownership of the different mobility tools leads to occurrence of more changes. This also applies to the simultaneous changes of the places of residence, education and employment, whereas moving out of the parents' house as well as the number of moves lower the risk. The number of persons born during the observed period as well as the household and accommodation size affects the duration positively. Respondents living abroad show considerably more variations. The index of purchasing power in the residential region decreases the durations.

Table 66   Hazard ratios of the combined duration models for the different types of durations (1985-2004)

Explanatory variable *(Average values for the observed durations)*	All durations without type	All durations with type	All durations with type as strata variable
Type of duration:			
Residential duration (referential category)			
Education duration		1.063	
Employment duration		1.316	
Car availability: Always		0.407	
Car availability: Partially		1.217	
National ticket ownership		1.260	
Regional ticket ownership		1.090	
Half-fare discount ticket ownership		0.860	
Left censoring of the duration	0.348	0.363	0.362
Age in years squared	0.999	0.999	0.999
Gender: Male	0.735	0.742	0.740
Age in years * Gender: Male	1.008	1.007	1.007
Nationality: Swiss national	1.102		
College or university degree	1.134	1.131	1.128
Share in education during the period			1.132
Duration in education at the beginning of the period in years	1.032	1.031	1.024
Changes in education during the period	0.771	0.759	0.744
Distance between the place of residence and the place of education in 1000 kilometres	1.220	1.220	1.242
Share in employment during the period	1.246		
Duration in employment at the beginning of the period in years		1.011	1.011
Changes in employment during the period	0.740	0.806	0.811
Distance between the place of residence and the place of employment in 1000 kilometres	1.137	1.123	1.119

Table 66 is continued ...

Table 66 continued ...

Explanatory variable (Average values for the observed durations)	All durations without type	All durations with type	All durations with type as strata variable
Monthly income in 1000 CHF	0.771	0.786	0.787
Monthly income in 1000 CHF squared	1.011	1.010	1.010
Monthly income natural logarithm	1.729	1.687	1.720
Car availability: Always		1.185	1.185
Car availability: Partially	1.238	1.245	1.231
National ticket ownership	1.604	1.613	1.620
Regional ticket ownership	1.172	1.160	1.175
Half-fare discount ticket ownership	1.095	1.157	1.159
Simultaneous change of the place of residence and the places of education or employment	1.670	1.627	1.624
Moving out of parents' house	0.819	0.843	0.835
Changes in residence during the period	0.620	0.619	0.622
Number of births in the household	0.619	0.629	0.638
Number of persons in the household	0.958	0.956	0.959
Number of rooms in the accommodation	0.947	0.943	0.938
Degree of urbanisation:			
Urban (referential category)			
Urban to rural			1.014
Rural			1.110
Place of residence abroad	1.469	1.370	1.363
Purchasing power index in the residential region	0.968	0.969	0.969
Number of observations	9565	9565	9565
Number of censored observations	3529	3529	3529
R^2 (generalised)	0.467	0.478	0.357

Furthermore, Table 67 presents the results of a multinomial logit model for the different duration types, using the residential durations as referential category. In this context, only uncensored cases, i.e., only the observations that are completely surveyed, are taken into account. The variance occurring between the different duration types is with approximately 80% very well explained. For all shown explanatory variables, the differences between the various types are significant. The estimated parameters mainly confirm the results of the previous models. For instance, the duration has a significantly negative effect on the different types, in comparison to the residential durations as referential category. This also applies to the occurrence of left censoring of the duration.

Table 67 Multinomial logit model for the different types of durations using the residential durations as referential category (1985-2004)

Explanatory variable	Education	Employment	Always available cars	Partially available cars	National tickets	Regional tickets	Half-fare discount tickets
Duration in years	− 0.066*	− 0.111*	− 0.062*	− 0.195*	− 0.190*	− 0.106*	− 0.114*
Left censoring of the duration	− 1.508*	− 1.405*	− 0.736*	− 1.997*	− 2.147*	− 1.464*	− 1.402*
Age in years	− 0.590*	− 1.137*	− 0.790*	− 1.252*	− 0.568	− 0.543*	− 0.734*
Age in years squared	+ 0.005*	+ 0.007*	+ 0.006*	+ 0.008*	+ 0.004	+ 0.004*	+ 0.006*
Age in years natural logarithm	+ 5.630*	+ 21.466*	+ 11.333*	+ 21.062*	+ 8.873	+ 7.043*	+ 9.977*
Nationality: Swiss national	+ 1.066*	+ 0.390*	+ 0.175	+ 0.450*	+ 0.914*	+ 0.119	+ 0.548*
College or university degree	+ 0.528*	− 0.398*	− 0.219	+ 0.163	+ 0.546*	− 0.042	+ 0.013
Duration in education at the beginning of the period in years	− 0.146*	− 0.123*	− 0.090*	+ 0.088*	− 0.087	− 0.092*	− 0.044
Changes in education during the period	− 5.648*	+ 0.293*	− 0.152	− 0.083	+ 0.065	+ 0.119	+ 0.083
Distance between the place of residence and the place of education in 1000 kilometres	+ 1.944*	− 34.212*	− 15.011*	− 4.406*	+ 1.525*	− 2.330	+ 1.403*
Duration in employment at the beginning of the period in years	− 0.044*	− 0.086*	+ 0.015	+ 0.046	+ 0.015	− 0.008	− 0.011
Changes in employment during the period	− 0.046	− 4.513*	+ 0.318*	+ 0.288*	− 0.005	+ 0.286*	+ 0.213*
Distance between the place of residence and the place of employment in 1000 kilometres	− 24.473*	+ 1.057*	− 14.081*	− 13.002*	− 4.932*	− 19.868*	− 17.949*
Monthly income in 1000 CHF	− 0.719*	+ 0.156*	+ 0.333*	− 0.464*	− 0.406*	− 0.348*	− 0.184*
Monthly income in 1000 CHF squared	+ 0.040*	− 0.009*	− 0.037*	+ 0.024*	+ 0.023*	+ 0.019*	+ 0.007
Simultaneous change of the places of residence and the places of education or employment	− 5.141*	− 5.038*	− 6.540*	− 6.778*	− 6.877*	− 6.756*	− 6.948*
Moving out of parents' house	− 3.224*	− 3.894*	− 3.208*	− 3.775*	− 3.231*	− 3.493*	− 3.536*
Duration in residence at the beginning of the period in years	+ 0.161*	+ 0.168*	+ 0.110*	+ 0.165*	+ 0.137*	+ 0.135*	+ 0.143*
Changes in residence during the period	+ 14.439*	+ 14.412*	+ 14.516*	+ 14.636*	+ 14.805*	+ 14.547*	+ 14.731*
Number of births in the household	− 1.115*	− 0.650*	− 0.509*	+ 0.009	− 0.862*	− 0.689*	− 0.694*
Number of persons in the household	− 0.006	− 0.195*	− 0.185	+ 0.095	− 0.090	+ 0.014	+ 0.075
Number of rooms in the accommodation	+ 0.288*	+ 0.272*	+ 0.218*	+ 0.199*	+ 0.306*	+ 0.247*	+ 0.262*
Degree of urbanisation: Urban (referential category)							
Urban to rural	− 0.066	− 0.078	− 0.091	+ 0.440*	− 0.717*	− 0.264	− 0.325*
Rural	− 0.301	+ 0.083	+ 0.360	+ 0.435*	− 1.137*	− 0.664*	− 0.609*
Place of residence abroad	− 0.424	− 0.679*	− 0.429	− 0.074	− 4.550*	− 1.617*	− 1.652*

Table 67 is continued ...

Table 67 continued ...

Explanatory variable	Education	Employment	Always available cars	Partially available cars	National tickets	Regional tickets	Half-fare discount tickets
Purchasing power index in the res. region	−0.011*	−0.018*	+0.002	−0.024*	+0.021*	−0.013*	−0.006
Constant	−7.034*	−44.297*	−24.809*	−41.993*	−21.945*	−12.852*	−20.618*

Number of observations	6036
R^2 (Cox and Snell)	0.787

* Level of significance ≤ 0.10

8.5 Changes in long-term and mid-term mobility

In this chapter, the changes in the places of residence, education and employment as well as in car availability and public transport season ticket ownership are analysed. In Table 68 the shares of these changes occurring within the same year are shown depending on the non-occurrence and occurrence of alterations in residence, education and employment. In the case of such a spatial change taking place, the share of another change is significantly higher, as a corresponding variance analysis indicates. This especially applies to the changes of the places of occupation.

Table 68   Changes in the places of residence, education and employment within the same year (1985-2004)

Change in ...	Place of residence		Place of education		Place of employment	
	No	Yes	No	Yes	No	Yes
Place of residence	0.0%	100.0%	13.2%	35.7%	11.5%	40.7%
Place of education	5.6%	17.8%	0.0%	100.0%	4.4%	27.4%
Place of employment	8.1%	31.7%	8.5%	43.2%	0.0%	100.0%
Car availability: Always	2.1%	9.6%	2.6%	10.1%	2.1%	10.8%
Car availability: Partially	2.1%	9.0%	2.2%	11.5%	2.2%	8.8%
National ticket ownership	1.3%	4.5%	1.2%	7.6%	1.3%	4.7%
Regional ticket ownership	2.6%	9.1%	2.3%	16.5%	2.2%	12.4%
Half-fare discount ticket ownership	3.2%	9.8%	3.2%	14.0%	3.1%	11.1%
Number of observations	17610	3080	21231	1569	20322	2478

Table 69 illustrates the correlations between the different types of changes. All correlation coefficients are significant at a 0.10 level. The strongest connections are observed for always and partially available cars as well as among the various public transport season tickets. The changes in residence, education and employment are also considerably related to one another, confirming findings of Rouwendal and van der Vlist (2005), as well as of others. Furthermore, education and the ownership of national and regional tickets show a relatively high association concerning the corresponding variations occurring within the same year.

Table 69 Correlations between all changes within the same year (1985-2004)

Change in …	Residence	Education	Employment	Always available cars	Partially available cars	National tickets	Regional tickets	Half-fare discount tickets
Place of residence	1.000	0.165	0.263	0.152	0.140	0.089	0.125	0.117
Place of education		1.000	0.283	0.111	0.140	0.126	0.202	0.140
Place of employment			1.000	0.157	0.124	0.084	0.180	0.128
Car availability: Always				1.000	0.437	0.077	0.124	0.111
Car availability: Partially					1.000	0.080	0.110	0.094
National ticket ownership						1.000	0.177	0.351
Regional ticket ownership							1.000	0.214
Half-fare discount ticket ownership								1.000
Number of observations	20690	22800	22800	22800	22800	22800	22800	22800

Figure 36 and Figure 37 show the alterations in the places of residence, education and employment during the observed time period from 1985 to 2004 and during the life course, respectively. In the two figures five years are grouped together. Over time the share of moves continuously increases, reaching a maximum of nearly 13% in the year 2000. This also applies at a slightly lower level to the changes in employment, whereas the share of changes in education is with about 4% relatively stable between 1985 and 2004. Most moves occur between the ages of 20 and 35 years, with a maximum of about 15%. Afterwards, the share of moves gradually decreases. This supports findings of Birg and Flöthmann (1992), Wagner (1990), and others. For the changes in the place of employment, the curve is again very similar at a lower level. Between the ages of 60 and 65 years the influence of retirement becomes visible. Variations in education occur, concurrent with the expectations, earlier during the life course. This share reaches a maximum for persons aged from 15 to 20 years.

Figure 36   Changes in residence, education and employment by time (1985-2004)

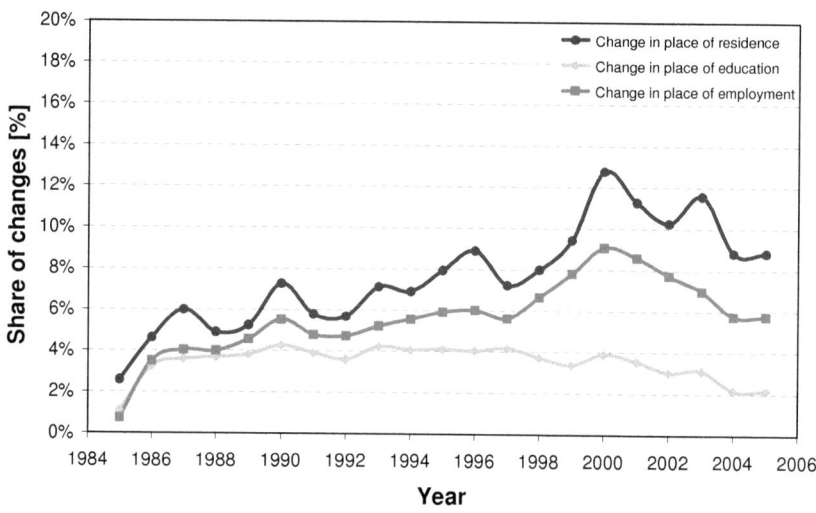

Figure 37   Changes in residence, education and employment by age (1985-2004)

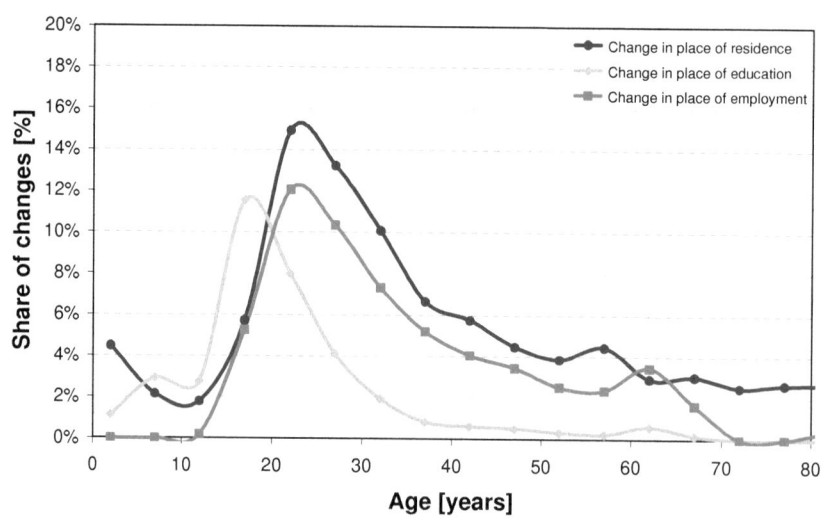

Figure 38 and Figure 39 present the changes in the ownership of the different mobility tools, once over time and once in regard to the age of the respondents. In comparison to the spatial changes, the shape of the curves is overall very similar, but ranging only up to 5% instead of up to 20%. In the course of time more variations are noticeable, with the least changes occurring in national ticket ownership and the most changes occurring in half-fare discount ticket ownership. Comparing the alterations in the ownership of always and partially available cars during the life course, the two maxima are slightly offset from one another, with always following partially car availability. After the age of 40 years, both curves become flat. There are some persons who give up their car, as they get older, but this happens only to a lesser extent. For the national tickets, the share of variations is noticeably lower, with the highest values being surveyed between the ages of 15 and 30 years. Regional tickets behave very similar to the partially available cars with a maximum for persons aged around 18 years. The half-fare discount tickets show larger variations with increasing age compared to the other mobility tools.

Figure 38   Changes in car availability and public transport season ticket ownership by time (1985-2004)

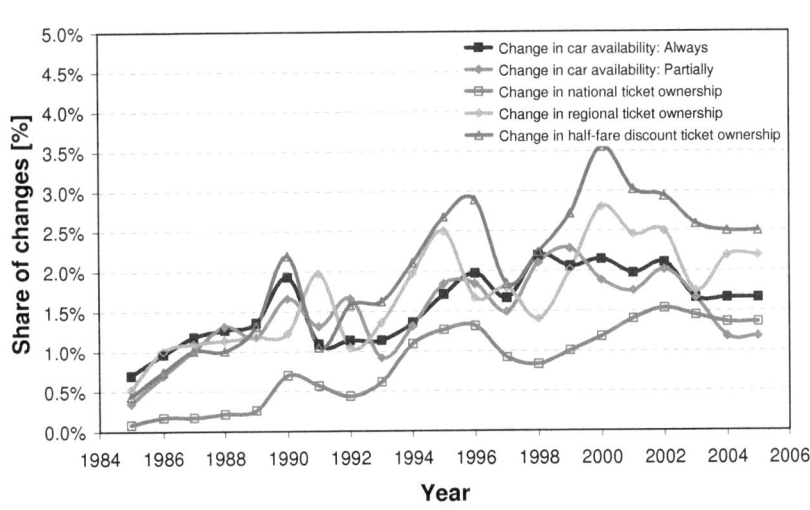

Figure 39   Changes in car availability and public transport season ticket ownership by age
(1985-2004)

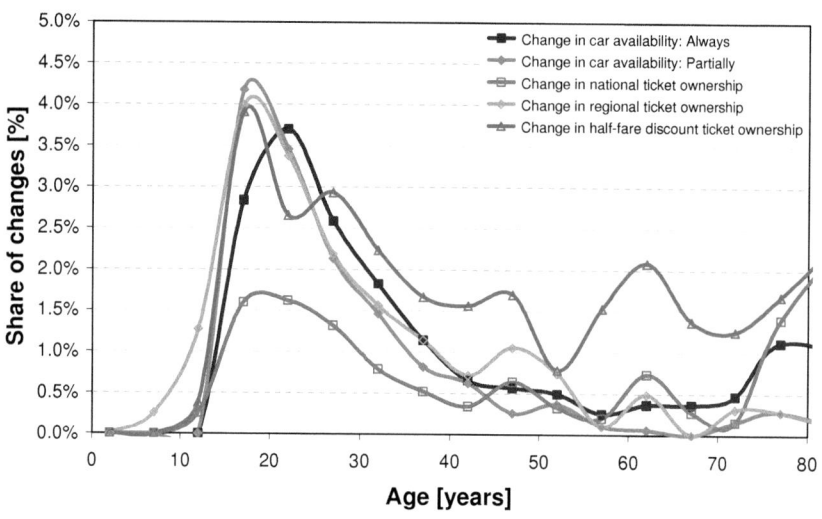

Table 70 shows the results of binomial logit models for the occurrence or non-occurrence of changes in residence, education and employment. Unfortunately, it is not possible to take into account the direction of the changes, i.e., starting or ending education and employment, since the proportion of changes in the data set is not sufficient to be further distinguished. For the explanatory variables used in the models, the difference between after and before each point in time is calculated on a semi-annual basis. Concerning the changes in residence, the probability increases with increasing age until reaching a maximum for persons aged between 25 and 30 years, and afterwards the propensity to move declines. This also applies to the alterations in employment. Changes in education are less likely to happen the older the respondents are. Overall, men show a more stable behaviour regarding the spatial changes than women. A rising income has a positive effect on the occurrence of all changes. Persons moving out of their parents' house tend to vary the places of education and employment more frequently at the same time. The birth of a person in the household leads to a higher propensity to move as well as to alter employment, confirming results by Aufhauser (1995). An increase in the household size influences moving in a negative way, whereas an increase in the accommodation size is related to more changes occurring. For respondents moving abroad, the probability of alterations taking place is reduced, contrary to the expectations. This is related to the specification of the corresponding influencing variable, which further distinguishes the direction of the moves. In this context, the share of persons moving to Switzerland is more than twice as high as the share of persons moving from Switzerland, thereby overemphasising the negative section of the variable. The panel effect is only relevant for variations in education and employment.

Table 70 Binomial logit models for the changes in residence, education and employment (1985-2004)

Explanatory variable *(Difference between the after and before status)*	Change in residence	Change in education	Change in employment	Change in education and employment
Age in years	+ 0.059 *	– 0.009	+ 0.076 *	+ 0.023 *
Age in years squared	– 0.001 *	– 0.001	– 0.001 *	– 0.001 *
Gender: Male	– 0.100 *	– 0.212 *	– 0.151 *	– 0.145 *
Nationality: Swiss national	– 0.011	+ 0.447 *	+ 0.107	+ 0.193 *
College or university degree	+ 0.060	+ 0.479 *	– 0.016	+ 0.102 *
Increase in distance between place of residence and place of education in 1000 kilometres	– 0.114	– 0.238 *	– 0.089	– 0.177
Increase in distance between place of residence and place of employment in 1000 kilometres	– 0.068	+ 0.496 *	– 0.046	– 0.043
Increase in monthly income in 1000 CHF	+ 0.089 *	+ 0.268 *	+ 0.017	– 0.050
Increase in monthly income in 1000 CHF squared	+ 0.027 *	+ 0.067 *	+ 0.163 *	+ 0.183 *
Moving out of parents' house		+ 1.640 *	+ 1.018 *	+ 1.358 *
Birth of a person in the household	+ 1.977 *	– 0.379	+ 0.588 *	+ 0.392 *
Increase in number of persons in household	– 1.406 *	– 0.017	– 0.104 *	– 0.041
Increase in number of rooms in accommodation	+ 0.599 *	+ 0.007	+ 0.038	+ 0.009
Increase in degree of urbanisation (from urban to rural)	– 0.180	+ 0.324 *	+ 0.081	+ 0.147
Increase in population in res. municipality in 1000 inhabitants	+ 0.013	+ 0.000	+ 0.001	+ 0.001
Increase in population density in res. municipality in 1000 inhabitants per square kilometre	+ 0.001 *	+ 0.185 *	– 0.004	+ 0.096
Change in place of residence from or to abroad considering the direction of the move (–1 = from abroad and +1 = to abroad)	– 3.161 *	– 0.926 *	– 1.989 *	– 1.741 *
Increase in purchasing power index in res. region	+ 0.001	– 0.013	+ 0.028 *	+ 0.017 *
Constant	– 3.408 *	– 2.761 *	– 3.826 *	– 2.593 *
Individual-specific random parameter	– 0.018	+ 0.558 *	– 0.537 *	– 0.493 *
Number of persons	1045	1045	1045	1045
Number of observations	31695	31695	31695	31695
ρ^2 (adjusted)	0.675	0.830	0.709	0.655

* Level of significance ≤ 0.10

Table 71 gives the estimated parameters of binomial models for the changes in mobility tool ownership. The age as well as the gender of the respondents has overall a negative influence on variations in car availability and public transport season ticket ownership. Swiss nationals tend to alter their ownership of mobility tools more frequently than foreign nationals. This also applies to persons holding a college or university degree, except for always available cars. An increase in the distance between the place of residence and the place of education decreases the probability of changes in car availability happening, whereas a place of employment which is further away leads to more alterations. The monthly income has again a positive effect. The move out of the parents' house results in the occurrence of more changes in the ownership of mobility tools. When a person is born in the household, changes in car availability become more likely, while, at the same time, a growth in household size reduces the respective probability. A move from or to another country leads, contrary to the expectations, to less alterations taking place. An increasing index of purchasing power in the residential region has a positive effect, especially for the variations occurring in public transport season ticket ownership.

Table 71 Binomial logit models for the changes in car availability and public transport season ticket ownership (1985-2004)

Explanatory variable *(Difference between the after and before status)*	Change in car availability: Always	Change in car availability: Partially	Change in national ticket ownership	Change in regional ticket ownership	Change in half-fare discount ticket ownership
Age in years	+ 0.064 *	+ 0.084 *	– 0.019	+ 0.021	– 0.008
Age in years squared	– 0.002 *	– 0.002 *	– 0.000	– 0.001 *	– 0.000
Gender: Male	– 0.187 *	– 0.327 *	– 0.313 *	– 0.364 *	– 0.277 *
Nationality: Swiss national	+ 0.030	+ 0.289 *	+ 0.731 *	+ 0.149	+ 0.358 *
College or university degree	– 0.019	+ 0.315 *	+ 1.076 *	+ 0.004	+ 0.457 *
Increase in distance between place of residence and place of education in 1000 kilometres	– 0.413 *	– 0.155	– 0.068	+ 0.105	+ 0.031
Increase in distance between place of residence and place of employment in 1000 kilometres	+ 0.051 *	+ 0.066 *	+ 0.087	+ 0.277 *	+ 0.080
Increase in monthly income in 1000 CHF	+ 0.257 *	+ 0.335 *	+ 0.041	+ 0.118 *	+ 0.089 *
Increase in monthly income in 1000 CHF squared	+ 0.008	– 0.004	+ 0.034 *	+ 0.028 *	+ 0.034 *
Moving out of parents' house	+ 1.597 *	+ 1.815 *	+ 1.293 *	+ 1.392 *	+ 0.947 *
Birth of a person in the household	+ 0.816 *	+ 0.759 *	– 0.641	+ 0.398	+ 0.445
Increase in number of persons in household	– 0.242 *	– 0.141 *	– 0.075	– 0.045	– 0.169 *
Increase in number of rooms in accommodation	+ 0.061	+ 0.048	+ 0.089	+ 0.120 *	– 0.002

Table 71 is continued ...

Table 71 continued ...

Explanatory variable *(Difference between the after and before status)*	Change in car availability: Always	Change in car availability: Partially	Change in national ticket ownership	Change in regional ticket ownership	Change in half-fare discount ticket ownership
Increase in degree of urbanisation (from urban to rural)	− 0.203	− 0.325	+ 0.293	− 0.245	+ 0.171
Increase in population in res. municipality in 1000 inhabitants	+ 0.001	+ 0.000	− 0.000	+ 0.001	+ 0.001 *
Increase in population density in res. municipality in 1000 inhabitants per square kilometre	− 0.025	− 0.051	+ 0.166	− 0.079	+ 0.002
Change in place of residence from or to abroad considering the direction of the move (−1 = from abroad and +1 = to abroad)	− 1.132 *	− 1.753 *	− 0.929	− 0.397	− 2.658 *
Increase in purchasing power index in res. region	+ 0.014	+ 0.023	+ 0.006	+ 0.046 *	+ 0.053 *
Constant	− 4.645 *	− 4.817 *	− 5.681 *	− 3.827 *	− 4.175 *
Individual-specific random parameter	+ 0.188	− 0.422 *	− 1.032 *	− 0.528 *	− 0.202
Number of persons	1045	1045	1045	1045	1045
Number of observations	31695	31695	31695	31695	31695
ρ^2 (adjusted)	0.903	0.898	0.928	0.883	0.861

* Level of significance ≤ 0.10

Figure 40 shows the distribution of the delays following a change in the place of residence until the next change in the places of education and employment, as well as vice versa. Around 50% of all changes are connected to a change in residence, education and employment within the first year. After that, the shares of the longer delays observed strongly decrease.

Figure 40 Distribution of the delays following a move until the next change in the places of education and employment, and vice versa (1985-2004)

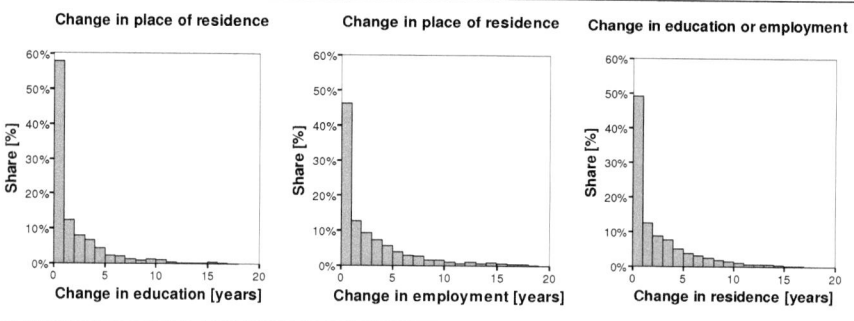

Figure 41 and Figure 42 illustrate the delays until the next change in car availability and public transport season ticket ownership following a move and a change in education or employment, respectively. These distributions look comparatively similar to the delays until the next variation in residence, education and employment, but the shares are lower overall. Respondents with always available cars show the most stable behaviour. In this group changes after a change in residence, education or employment occur for only about 20% of the persons within the first year, whereas this share amounts to about 30% to 35% for persons with partially available cars. For the national and regional tickets, approximately one third of all the delays are shorter than one year. The changes in half-fare discount ticket ownership show trends comparable to the always available cars. And again, the shares of the longer durations until the next change in mobility tool ownership decrease strongly after the first year.

Figure 41 Distribution of the delays following a move until the next change in car availability and public transport season ticket ownership (1985-2004)

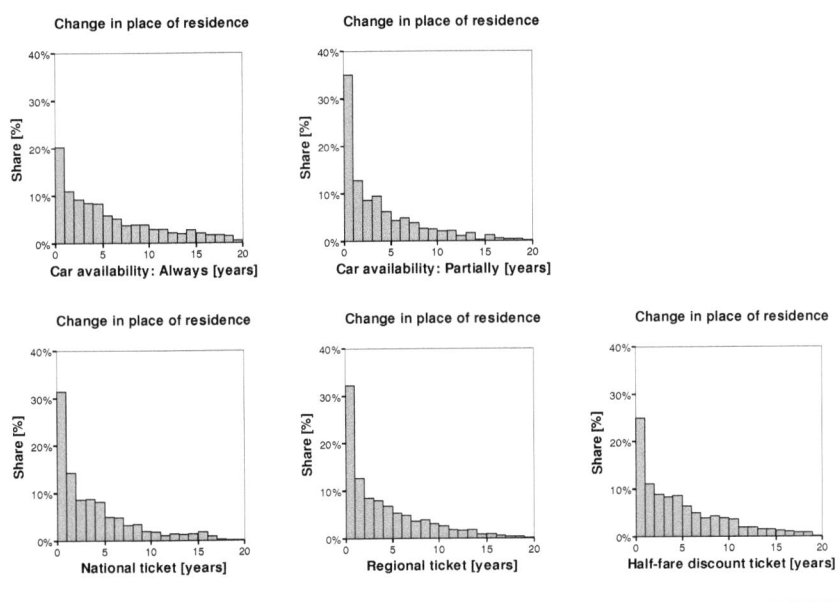

Figure 42 Distribution of the delays following a change in education or employment until the next change in car availability and public transport season ticket ownership (1985-2004)

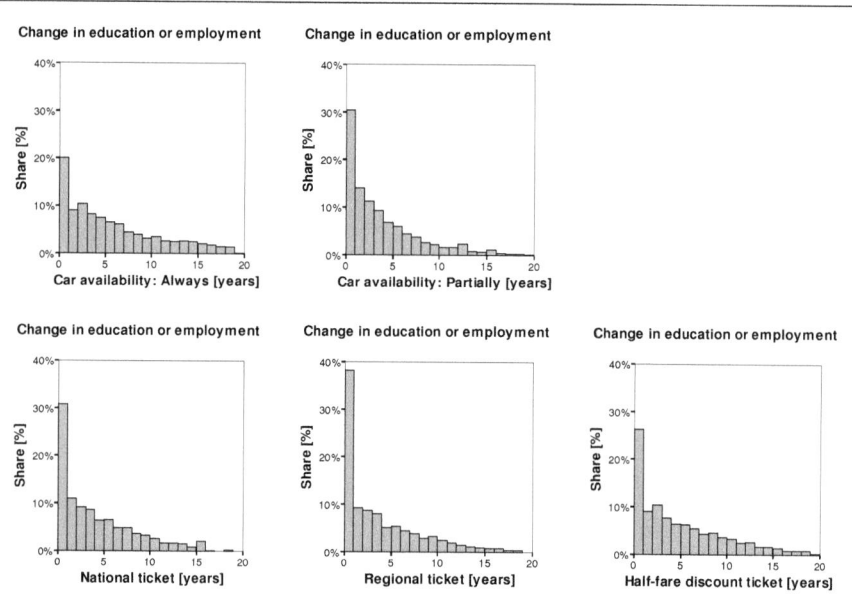

An analogue picture arises for the distribution of these durations after variations in car availability and public transport season ticket ownership. In this context, changes among the different mobility tools are very strongly connected to one another.

In the following, the delays occurring until the next changes in the ownership of mobility tools are analysed, firstly subsequent to a move and secondly subsequent to a change in education or employment. In this context, Figure 43 and Figure 44 illustrate the corresponding hazard rates. Both graphs look relatively similar. In comparison to the general durations of the mobility tool ownership shown in Figure 33, the hazard rates are to some extent higher. For the always available cars, the curves are the flattest ones, followed by the half-fare discount tickets. The partially available cars as well as national and regional tickets show a clear increase for the first year, after which the hazard rates decrease. Regarding the longer durations, greater variations occur, since these estimates are only based on a relatively small number of observations.

Figure 43 Hazard rates for the delays following a move until the next change in car availability and public transport season ticket ownership (1985-2004)

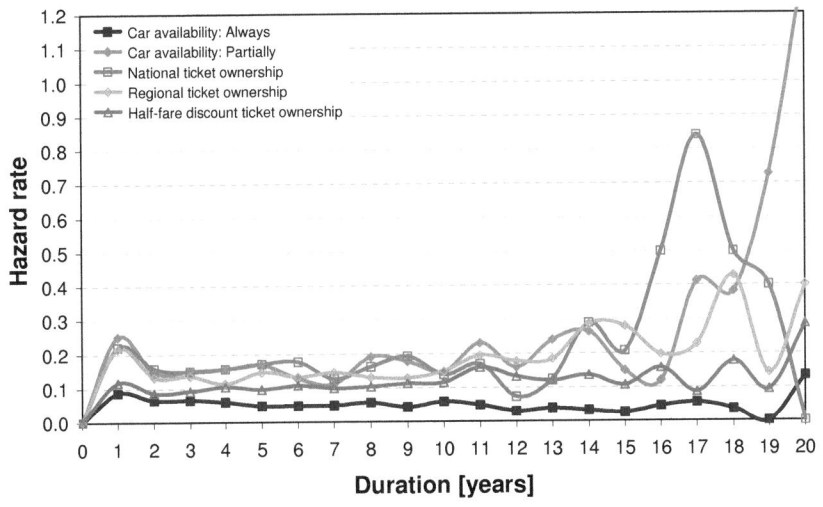

Figure 44 Hazard rates of the delays following a change in education or employment until the next change in car availability and public transport season ticket ownership (1985-2004)

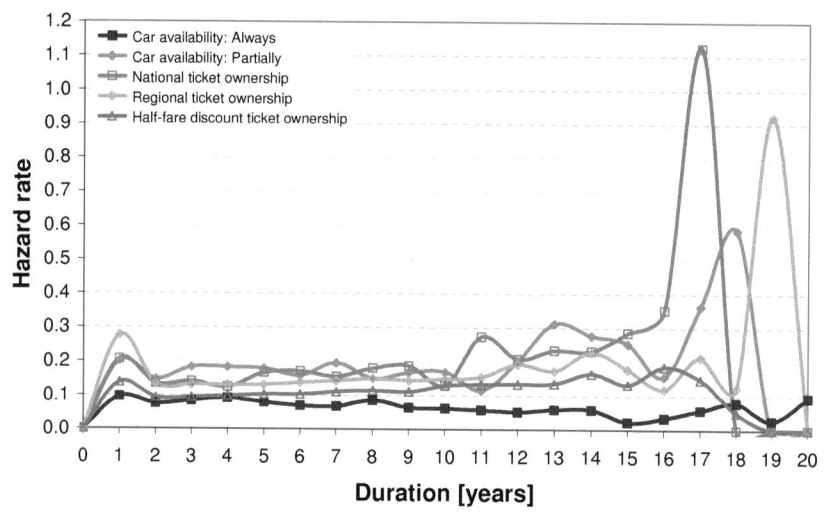

Table 72 and Table 73 list the hazard ratios for the delays until the next change in car availability and public transport season ticket ownership after moving as well as after changing education or employment. Again, the results for the two situations are quite similar. With increasing age the hazard decreases, especially after reaching the age of 30 years. Men are in general at a lower risk than women, in the cases where gender has an effect. Swiss nationals tend to alter their ownership of half-fare discount tickets later after a change. For car availability, the duration following a move is negatively influenced by a college or university degree. The variables describing education and employment only play a minor role for some of the mobility tools. Higher fuel prices lead to shorter durations concerning always available cars and half-fare discount tickets. Already having a car at disposal at the point of time, when a change in residence, education or employment occurs decreases, the probability of changes in car availability of the equal level, whereas an always available car increases the hazard for partially available cars, and vice versa. This means that cars are acquired rather than abandoned, confirming findings of Dargay (2001), providing a further indication for the stability of car availability. Among the various public transport season tickets this relationship between the ownership of the same and another type exists as well. At the same time, car availability has only an influence for the half-fare discount ticket ownership, which is positive. A simultaneous change of the place of residence and the places of education or employment leads to a shorter duration until the next alteration in mobility tool ownership. The number of rooms in the accommodation diminishes the probability of such changes occurring. With each additional room the hazard rate declines by 5% to 9%. The index of purchasing power in the residential region has a negative effect on the delays, concurrent with the expectations.

Table 72 Hazard ratios of the duration models for the delays following a move until the next change in car availability and public transport season ticket ownership (1985-2004)

Explanatory variable *(Values at the time of moving)*	Change in car availability: Always	Change in car availability: Partially	Change in national ticket ownership	Change in regional ticket ownership	Change in half-fare discount ticket ownership
Age in years		1.111		0.946	
Age in years squared		0.998	1.000		1.000
Age in years natural logarithm				4.551	1.780
Gender: Male	1.526	0.746			
Age in years * Gender: Male	0.979				
Nationality: Swiss national					0.754
College or university degree	1.203	1.249			
Duration in education at the beginning of the period in years	1.049			0.965	1.066
Changes in education during the period		0.887			
Fuel price in 0.01 CHF per litre (lead free 95)	1.027				1.024
Car availability: Always	0.089	1.805			0.825
Car availability: Partially	1.269	0.415			
National ticket ownership					1.776
Regional ticket ownership	1.221		1.586	0.383	
Half-fare discount ticket ownership			1.696		0.240
Simultaneous change of the place of residence and the places of education or employment	1.290	1.303		1.251	1.198
Moving out of parents' house				1.195	
Number of rooms in the accommodation	0.911		0.946		
Place of residence abroad		0.724			
Purchasing power index in the residential region		1.025	1.042	1.025	
Number of observations	1967	846	524	1029	1621
Number of censored observations	1318	231	132	329	764
R^2 (generalised)	0.383	0.237	0.163	0.190	0.264

Table 73 Hazard ratios of the duration models for the delays following a change in education or employment until the next change in car availability and public transport season ticket ownership (1985-2004)

Explanatory variable (Values at the time of changing education or employment)	Change in car availability: Always	Change in car availability: Partially	Change in national ticket ownership	Change in regional ticket ownership	Change in half-fare discount ticket ownership
Age in years	1.093	1.114		1.075	1.040
Age in years squared	0.998	0.998	1.000	0.999	0.999
Age in years * Gender: Male	0.994				
Nationality: Swiss national					0.817
Changes in education during the period	1.340				
Distance between the place of residence and the place of employment in 1000 kilometres			3.124		
Monthly income in 1000 CHF	0.955				
Fuel price in CHF per litre (lead free 95)	1.029				1.024
Car availability: Always	0.135	1.871			0.804
Car availability: Partially		0.496			
National ticket ownership			0.714		1.422
Regional ticket ownership			1.288	0.515	
Half-fare discount ticket ownership	1.171		1.342		0.301
Simultaneous change of the place of residence and the places of education or employment				1.436	
Duration in residence at the beginning of the period in years				1.013	
Changes in residence during the period				0.838	
Number of births in the household			0.726		0.774
Number of rooms in the accommodation	0.945				
Purchasing power index in the residential region		1.031	1.048	1.017	
Number of observations	1981	1121	665	1306	1805
Number of censored observations	1081	268	140	362	728
R^2 (generalised)	0.390	0.205	0.170	0.111	0.219

Further analyses can include the closer examination of other events, e.g., marriage, divorce, the birth of children in a family, etc. (Lanzendorf, 2006; Prillwitz et al., 2006). In this context, both durations before and after another event occurs can be analysed.

9 Summary of the analyses and conclusions

In the following, the results of the various analyses are summarised with respect to the research questions stated in the fourth chapter. Based on these results, corresponding conclusions are drawn as well as implications for policy and planning are specified.

9.1 Summary of the analyses

Using the data collected by means of a retrospective survey covering the 20 year period from 1985 to 2004, analyses over time and over the life course, concerning the residential and occupational behaviour on the one hand and the ownership of mobility tools on the other hand, are carried out. Furthermore, the various durations and occurring changes in long-term and mid-term mobility are described.

Developments over time and over the life course

Over time, the occurrence of important personal and familial events, such as the move out of the parents' house, the birth of persons in the household, partnerships and marriages as well as break-ups and divorces, is fairly constant between 1985 and 2004, with shares ranging between 0% and 2%. The share of moves continuously increases over time, reaching a maximum of nearly 13% in the year 2000. This also applies at a slightly lower level to the changes in employment, whereas the share of changes in education is with about 4% relatively stable. At the beginning of the observed period nearly one third of the persons are in education and employment. Over time, as people get older, persons in education diminish till approximately 9% in the year 2005, whereas the share of employed persons increases up to 72%. Accordingly, the income per month rises as well by 87% from a level of about 3100 CHF in the year 1985 to a level of about 5700 CHF in the year 2005. The median distances between the place of residence and the corresponding places of education and employment increase over time, to a greater extent for education than for employment. During the observed 20 year period major changes concerning the main mode of transport to the place of education occur. There is a small increase in the share of private transport by about 10% and a much bigger increase in the share of public transport by over 30%, whereas cycling and walking strongly decline. By contrast, the main mode for the trip to the place of employment remains relatively stable over time. Approximately 50% of the respondents use private transport, 30% public transport and nearly 10% each cycle and walk. These developments are closely connected to the changes in the median distance from home to the places of occupation. During the period from 1985 to 2004 the ownership of all mobility tools increases. The availability of only a car declines over time, whereas the share of car and public transport season ticket owners increases from 20% to 45%. At the same time, respondents without any mobility tools diminish during the observed time period. In comparison to the spatial changes, the changes in the ownership of the different mobility tools are overall very similar, but ranging only up to 5% instead of up to 20%. Over the course of time more variations are noticeable, with the least changes occurring in national ticket ownership and the most changes occurring in half-fare discount ticket ownership. Overall, these developments over time are consistent with the composition and the expected developments of the sample.

Over the life course, the most personal and familial events occur rather early, consistent with the literature (Aufhauser, 1995; Birg and Flöthmann, 1992; Wagner, 1990). The move out of the parents' house primarily takes place at the age between 15 and 20 years. The birth of persons in the household shows a maximum at the beginning, covering the respondents' birth as well as the birth of siblings. Between the ages of 20 and 30 years again a higher number of children are born, followed by a gradual decrease afterwards. The share for the formation of partnerships and marriages is highest for persons aged from 15 to 30 years. Break-ups and divorces show overall relatively low shares with values not exceeding 1%. Most moves occur between the ages of 20 and 35 years, with a maximum of about 15%. Afterwards, the share of moves gradually decreases. This supports findings of Birg and Flöthmann (1992), Wagner (1990), and others. The changes in the place of employment are again very similar at a lower level. Between the ages of 60 and 65 years the influence of retirement becomes visible. Variations in education occur, concurrent with the expectations, earlier during the life course. This share is highest for persons aged from 15 to 20 years. The share of persons in education reaches a maximum at the age of about 16 years. Afterwards, it strongly decreases until the age of about 32 years. Employment shows a clear increase between the ages of 15 and 30 years, followed by a relative stable period with a share of about 80% until persons start retiring, when they reach the age of circa 60 years. The monthly income continuously increases over the life course, especially for persons aged from 15 to 30 years. Only after the age of 65 years a small reduction is observable. The median distance to the place of education shows a strong rise between the ages of 10 and 18 years. With increasing age the distance to education fluctuates, which is connected to more specialised educations and a rather low number of observed cases for the older age groups. For the distance to the place of employment, there is likewise a strong increase noticeable until the age of 18 years. Then, a slow decrease until the retirement age and an afterwards stable section with a median distance of about 1 kilometre follow. Concerning the mode of transport most frequently used for the trip from home to the places of occupation, for persons in education, there is a strong increase in the usage of private and public transport observable, whereas walking is only until the age of about twelve years of some importance. For the trip to the place of employment, persons aged from 25 to 65 years show a relative stable modal split. These trends are again strongly related to the changes in the median distance between the place of residence and the corresponding places of education and employment. Regarding the age of the respondents, there is, as expected, a strong increase in car ownership after reaching the age of 18 years. Persons aged from 25 to 50 years show the highest share with about 75%. Then, a slow decrease is noticeable. The ownership of national tickets increases over the life course, whereas the share of regional tickets decreases. The half-fare discount tickets have growing shares. About one third of the respondents own a car and public transport season tickets at the same time. Overall, the ownership of mobility tools increases at the beginning and then remains relatively stable over the life course with only approximately 10% of persons not having any mobility tool at their disposal. Comparing the alterations in the ownership of always and partially available cars during the life course the, two maxima at the ages of about 23 and 18 years, respectively, are slightly offset from one another, with always following partially car availability. After the age of 40 years, the shares become lower. There are some persons who give up their car, as they get older, but this happens only to a lesser extent. For the national tickets, the share of variations is noticeably lower, with the highest values being surveyed between the ages of 15 and 30 years. Regional tickets behave very similar to the partially available cars with a maximum for persons aged around 18 years. The half-fare discount tickets show larger variations with increasing age compared to the other mobility tools.

Places of residence and moving behaviour

Concerning the moving behaviour, the observed residential durations for the period from 1985 to 2004 are on average about 7 years long with a standard deviation of about 9 years. Approximately two thirds of the durations are up to five years, nearly 80% up to ten years and nearly 90% up to twenty years long.

Overall, a place of residence is approximately 481 kilometres away from the previous one with a standard deviation of 2109 kilometres. The distribution of the distances between two successive places of residence is very strongly left-skewed. About 90% of all residential distances lie in the range from 0 kilometres to 250 kilometres. Over one third of all the moves take place within a radius of 5 kilometres. This confirms the statement that most residential moves are characterised by short distances (Blijie, 2005; Franz, 1984). The observed distances are smallest in the rural areas and by contrast considerably above average, when persons move abroad, concurrent with the expectations.

The respondents tend to move to slightly bigger accommodations. This especially applies for the smaller accommodations with sizes ranging from one room to three rooms. Overall, the number of rooms before and after a move is positively correlated in a significant way.

Concerning the reasons for moving, personal and familial reasons with 41% are given in the first place, followed by accommodation related reasons with 26%. However, these two categories are closely connected to one another. In general, motives behind residential mobility are difficult to separate, and not easily to assign. In many cases several factors together lead to a move (Birg and Flöthmann, 1992). Education and employment related reasons have a share of about 26%. This reason is especially important for moving abroad, whereas other categories play a subordinate role for this kind of moves.

Ownership of mobility tools

Overall, 48% of the respondents have a car always and 11% partially at their disposal. Concerning public transport, 7% and 17% own a national or a regional season ticket, respectively, and 31% a half-fare discount ticket during the observed period of time from 1985 to 2004. Car availability and public transport season ticket ownership are lowest abroad. In Switzerland, the main centres also show low car ownership, whereas the share of season ticket owners is comparatively high, which is concurrent with the expectations and the literature (Beige, 2004; Karlaftis and Golias, 2002).

Furthermore, the ownership of mobility tools is analysed by gender, age and birth cohort membership. Thereby, significant differences occur for all three variables. For the ownership of cars, including always and partially available cars, it is noticeable that the oldest cohort group owns considerably fewer cars than the younger cohorts. Highest is the ownership among those who are 35 to 55 years old today. At the same time, men have noticeably more frequently a car at their disposal than women of the same age. However, for the younger generations this difference diminishes. Except for the oldest group, there are still increases in ownership observable within the different cohorts. This means that the level of saturation is not necessarily reached yet. In this context, further substantial increases in the ownership of cars are to be expected. Unlike the older generations of today, which still have a relatively low level of motorisation now, the older generations of tomorrow will continue owning their cars, as they age. Comparing the ownership of national and regional season tickets to car ownership, there is a different trend visible. The cohorts with a current age between 35 and 55 years now show the lowest ownership rates. Furthermore, women generally own more public transport season tickets than men. Therefore, both car ownership on the one side and national and regional season ticket ownership on the other side substitute one another. This is consistent with the findings of Simma and Axhausen (2003). The ownership of half-fare discount tickets increases relatively strongly over the life course. With the exception of the oldest cohort group, the female respondents tend to own more half-fare discount tickets than the male respondents of the same age.

Various discrete choice models are estimated for the ownership of mobility tools between 1985 and 2004. In this context, only persons that are 18 years and older are considered. The results of different binomial logit models for the availability of cars and the ownership of public transport season tickets show that the probability of disposing of a car which is available at all times increases until the age of 54 years and then slowly declines. At the same time, male respondents are more often in this position than female respondents. Employment as well as income has a positive influence. The ownership of season tickets is related to a lower proportion of always available cars. In larger households these cars are less likely, as available cars are more likely to be shared. With reference to the urban areas, persons having their own car live more often in rural areas. Rather opposed tendencies are noticeable for partially available cars. For instance, men tend to have a car less frequently only part time at their disposal than women. Furthermore, the monthly income has a negative effect. Persons owning public transport season tickets are in general more likely to simultaneously have a partially available car. The ownership of national and regional tickets for public transport is reduced with increasing age, while the utility for half-fare discount tickets also increases. Men tend to own less public transport season tickets than women. The Swiss nationality as well as a college or university degree and being in education lead to a higher ownership of these mobility tools. Only for the national tickets the distance to the place of employment has a positive influence. Concurrent with the expectations, simultaneous car availability decreases the ownership of public transport season tickets. Persons with national or regional tickets live more often in urban areas. In the cases where the place of residence is abroad, season ticket ownership tends to be lower. The index of purchasing power in the region of residence, measuring the changes in consumer prices in a country in Euro and making an adjustment for changes in exchange rates, has a positive influence on the public transport season ticket ownership. All of these results are in general consistent with other analyses concerning the ownership of mobility tools (Beige, 2004; Karlaftis and Golias, 2002; Simma and Axhausen, 2003). For the standard deviation of the individual-specific error term, which takes the panel effect into account, relatively high values are estimated. This indicates a substantial heterogeneity in the sample.

When considering the ownership of mobility tools in six different groups, which cover all possible combinations, the largest group with about one third are mere car owners, followed by persons having a car and public transport season tickets at their disposal. For persons living abroad, the share with no mobility tools is with over 30% considerably higher than for Switzerland, where the lowest shares are found in the middle and ancillary centres as well as in the agglomeration municipalities. Mere public transport season tickets owners tend to live in the main centres, while non-ownership is more widespread abroad, in rural areas and in agglomeration municipalities. The share of the group with a car, but no season tickets lies noticeably above the average in other countries, whereas in Switzerland persons tend to own a car and season tickets more frequently simultaneously. This is probably primarily connected to the prominent quality and quantity of the Swiss public transport systems. Concerning the changes in mobility tool ownership taking place during the period from 1985 to 2004, in about one third of all cases respondents acquire a car, whereas only 10% are related to the abandonment of a car. To a slightly lesser extent and in a more balanced way this also applies to the various public transport season tickets. Overall, the highest shares arise for the transitions to the alternatives including cars, thereby again implying a future increase in car ownership.

In the various discrete choice models estimated for the mobility tool ownership in six different groups, age has a positive effect with reference to the group of mere car owners, especially for the older respondents. Only the alternatives including national and regional season tickets show a slight decrease of the utility for younger persons. Gender has a negative influence. This means that men tend to merely own a car more frequently than women. Respondents without any mobility tools are more likely to be foreign national as well as not to hold a college or university degree. The opposite tendency is visible for the other groups which own mobility tools. For these groups, being in education, a change in education and the corresponding distance show a positive influence. Persons with no mobility tools or with only a half-fare discount ticket are less likely to be employed, whereas employment and a change in employment increases the probability of having a car and public transport season tickets simultaneously at disposal. At the same time, the distance between the place of residence and the place of employment has a negative effect for these last two groups. This means that persons which are more mobile, covering longer distances to their places of occupation, are also more likely to own mobility tools, and vice versa. Therefore, the statement that mobility tool ownership can be used as a proxy for the actual travel behaviour is supported (Prillwitz *et* al., 2006; Simma and Axhausen, 2003). With increasing income the propensity to not have a mobility tool or to not have a car decreases. A higher income enhances the simultaneous availability of cars and public transport season tickets, emphasising the cost dimension of the mobility tool ownership. The birth of a person in the household as well as the household size and accommodation size increase the ownership of a car. Mere car owners tend to live in more rural areas as well as not in Switzerland. Thereby, the index of purchasing power in the residential region has a negative influence on car ownership. The various logit models are relatively similar to one another regarding the significant influencing variables as well as the estimated parameters, with the best model being the nested logit model with the two nests for owning a car and not owning a car. This indicates that structuring the sample by the ownership and non-ownership of cars is most appropriate in comparison to the public transport season tickets, pointing to a stronger commitment towards cars. When the correlations between the different groups are incorporated in the probit model, these correlations are in all cases significant. In this context, the largest values are observed among the alternatives including a car. In comparison to the logit models, the goodness of fit measure improves considerably. At the same time, considerably fewer influencing variables are significant in the probit model than in the logit models. Their impact is captured by the correlations between the different mobility tool ownership groups.

Regarding the observed durations of car availability and public transport season ticket ownership, for about one third, cars are always available from 1985 to 2004. Partial car availability is more often indicated for shorter periods of time with over 50% being less than five years long and over 80% being less than ten years long. For the ownership of national and regional tickets, the highest shares occur for durations shorter than five years. To a lesser extent this also applies to the half-fare discount ticket ownership. Overall, the ownership of the different mobility tools is relatively stable over time, especially the availability of cars, whereas the slightly more variable ownership of season tickets during the period from 1985 to 2004 points to a weaker commitment to public transport. This stability in mobility tool ownership over longer periods of time is also found in other studies (Axhausen and Beige, 2003; Bjørner and Leth-Petersen, 2005; Lanzendorf, 2006; Prillwitz et al., 2006; Simma and Axhausen, 2003).

Long-term and mid-term mobility

With increasing age persons tend to move less and to stay longer at a place of residence. This result is consistent with most of the literature (Clark and Onaka, 1983; Courgeau, 1985; Vandersmissen et al., 2005). For example, Hollingworth and Miller (1996) found that age is the variable with the most distinct effect on residential mobility. The gender of the respondents has no influence. Respondents with a college or university degree are more likely to change their place of residence. Changes in education and employment during the observed period lead to a lower probability of moving, contradicting findings of Hollingworth and Miller (1996), Rouwendal and van der Vlist (2005), Van der Waerden *et al.* (2003), Verhoeven *et al.* (2005), and others. The share in education affects the duration in employment positively, and vice versa. The distances between the places of residence, education and employment increase the various hazards of changes occurring considerably. So, persons seem to compensate distances by shorter stays. This also supports the finding that changes in education and employment decrease the probability of moving, when the distances to the places of occupation are reduced in these cases. The ownership of the different mobility tools, especially the national and regional tickets, shows positive hazard parameters, indicating that more mobile persons are, at the same time, also more spatially mobile, concurrent with the expectations. Simultaneous changes of the places of residence, education and employment strongly increase the hazard of changes occurring, thereby confirming results in the literature (Lelièvre and Bonvalet, 1994). Moving out of the parents' house leads to a longer stay at the following place of residence. The number of moves has a positive effect on the durations of education and employment. The number of births in the household leads to longer durations between moves. With each birth the hazard rate decreases by about 30.9%. The household size and the accommodation size affect the various durations positively as well. This also applies to the education and employment durations. In the case that the place of residence is abroad, the observed durations tend to be considerably shorter. The index of purchasing power in the residential region has a hazard ratio that is smaller than one, thereby indicating an increase in the probability of changes occurring.

The mobility tool ownership durations are positively influenced by the age of the respondents, especially this is true for cars as well as for national and regional tickets, confirming the general hypothesis that over the life course younger adults tend to be more open to change than the elderly. Men are more likely to hold a national ticket over longer periods of time than women. Changes of the places of residence, education and employment during the observed period reduce the hazard significantly. Longer distances from the place of residence to the place of education lead to shorter durations of public transport season ticket ownership. Persons show with increasing income a higher stability for always available cars and a lower stability for partially available cars. Fuel prices affect the durations considerably in a positive way, quite contrary to the expectations. In this context, it is necessary to consider that the fuel prices used in the analyses are not spatially disaggregated, and, therefore, rather represent the development over time. The ownership of a half-fare discount ticket increases the hazard for always available cars. When a simultaneous change of residence, education and employment occurs, the ownership durations tend to be much shorter. This points to inter-regional migration rather than intra-regional migration taking place, i.e., all activity places change after a move (Franz, 1984; Scheiner, 2006). In this context, the actual travel behaviour seems to be reconsidered and altered, as the spatial-temporal configuration of the activity space shifts (Van der Waerden et al., 2003; Verhoeven et al., 2005). The same applies to the moves out of the parents' house and the effect on car availability. The number of births in the household reduces the probability of variations in mobility tool ownership. Concerning the degree of urbanisation, the durations for which persons own a half-fare discount ticket are significantly shorter in the urban areas than in the rural areas. With an increasing index of purchasing power the hazard rate for national tickets decreases.

Furthermore, the changes in the places of residence, education and employment as well as in car availability and public transport season ticket ownership are analysed. In the case of alterations in residence, education and employment, the share of another change is significantly higher. This especially applies to the changes of the places of occupation. The correlations between the different types of changes are all significant. The strongest connections are observed for always and partially available cars as well as among the various public transport season tickets. The changes in residence, education and employment are also considerably related to one another, confirming findings of Rouwendal and van der Vlist (2005), as well as of others. Furthermore, education and the ownership of national and regional tickets show a relatively high association concerning the corresponding variations occurring within the same year. This points to the important role of the public transport concerning the trip from home to the place of education, also found in the analyses of the most frequently mode of transport used to the places of occupation.

Concerning the changes in residence, the probability increases with increasing age until reaching a maximum for persons aged between 25 and 30 years, and afterwards the propensity to move declines. This also applies to the alterations in employment. Changes in education are less likely to happen the older the respondents are, as one expects. Overall, men show a more stable behaviour regarding the spatial changes than women. A rising income has a positive effect on the occurrence of all changes. Persons moving out of their parents' house tend to vary the places of education and employment more frequently at the same time. The birth of a person in the household leads to a higher propensity to move as well as to alter employment, confirming results by Aufhauser (1995). An increase in the household size influences moving in a negative way, whereas an increase in the accommodation size is related to more changes occurring. For respondents moving abroad, the probability of alterations taking place is reduced, contrary to the expectations. This is related to the specification of the corresponding influencing variable, which further distinguishes the direction of the moves. In this context, the share of persons moving to Switzerland is more than twice as high as the share of persons moving from Switzerland, due to the fact that the retrospective survey was carried out in Switzerland, thereby overemphasising the negative section of the variable.

The age as well as the gender of the respondents has overall a negative influence on variations in car availability and public transport season ticket ownership. Swiss nationals tend to alter their ownership of mobility tools more frequently than foreign nationals. This also applies to persons holding a college or university degree, except for always available cars. An increase in the distance between the place of residence and the place of education decreases the probability of changes in car availability happening, whereas a place of employment which is further away leads to more alterations. The monthly income has again a positive effect. The move out of the parents' house results in the occurrence of more changes in the ownership of mobility tools. When a person is born in the household, changes in car availability become more likely, while, at the same time, a growth in household size reduces the respective probability. A move from or to another country leads, contrary to the expectations, to less alterations taking place. An increasing index of purchasing power in the residential region has a positive effect, especially for the variations occurring in public transport season ticket ownership.

The analyses of the delays between changes in the places of residence, education and employment show that around 50% of all changes are connected to a change in residence, education and employment within the first year. After that, the shares of the longer delays observed strongly decrease. Concerning the delays until the next change in mobility tool ownership following a move and a change in occupation, the connection is to some extent weaker. Respondents with always available cars show the most stable behaviour. In this group changes after a change in residence, education or employment occur for only about 20% of the persons within the first year, whereas this share amounts to about 30% to 35% for persons with partially available cars. For the national and regional tickets, approximately one third of all the delays are shorter than one year. The changes in half-fare discount ticket ownership show trends comparable to the always available cars. And again, the shares of the longer durations until the next change in mobility tool ownership decrease strongly after the first year. An analogue picture arises for the distribution of these durations after variations in car availability and public transport season ticket ownership. In this context, changes among the different mobility tools are very strongly connected to one another.

Duration models for the delays until the next change in car availability and public transport season ticket ownership after moving as well as after changing education or employment indicate that with increasing age the hazard decreases, especially after reaching the age of 30 years. Men are in general at a lower risk than women, in the cases where gender has an effect. Swiss nationals tend to alter their ownership of half-fare discount tickets later after a change. For car availability, the duration following a move is negatively influenced by a college or university degree. The variables describing education and employment only play a minor role for some of the mobility tools. Higher fuel prices lead to shorter durations concerning always available cars and half-fare discount tickets. Already having a car at disposal at the point of time, when a change in residence, education or employment occurs, decreases the probability of changes in car availability of the equal level, whereas an always available car increases the hazard for partially available cars, and vice versa. This means that cars are acquired rather than abandoned, confirming findings of Dargay (2001), providing a further indication of the stability of car availability. Among the various public transport season tickets this relationship between the ownership of the same and another type exists as well. At the same time, car availability has only an influence for the half-fare discount ticket ownership, which is positive. A simultaneous change of the place of residence and the places of education or employment leads to a shorter duration until the next alteration in mobility tool ownership. The number of rooms in the accommodation diminishes the probability of such changes occurring. The index of purchasing power in the residential region has a negative effect on the delays, concurrent with the expectations.

9.2 Conclusions and implications for policy and planning

Based on the results of the various analyses, the following conclusions are drawn as well as implications for policy and planning are specified.

The analyses concerning long-term and mid-term mobility show that the ownership of the different mobility tools is relatively stable over time, especially the availability of cars, whereas the slightly more variable ownership of season tickets points to a weaker commitment to public transport. Concerning the changes in mobility tool ownership taking place during the period from 1985 to 2004, only about 3% of the respondents vary their mobility tool ownership each year. In about one third of all cases respondents acquire a car, whereas only 10% are related to the abandonment of a car. To a slightly lesser extent and in a more balanced way this also applies to the various public transport season tickets. In contrast, alterations in residence, education and employment occur noticeably more frequently, e.g., with about 15% of all the persons moving within each year. Approximately 70% of all the corresponding residential and occupational durations observed during the period from 1985 to 2004 are up to five years long. Overall, spatial changes as well as changes in mobility tool ownership are considerably connected to one another, this means that changes in the different dimensions of life tend to occur simultaneously. Around 50% of all spatial changes are related to a change in residence, education and employment within the first year. After that, the shares of the longer delays observed strongly decrease. Concerning the delays until the next variation in mobility tool ownership following a move and a change in occupation, the connection is to some extent weaker. In this context, changes among the different mobility tools are very strongly connected to one another.

In summary, one can say that there exists a strong interrelation between the two examined aspects of long-term and mid-term mobility. The residential mobility is influenced by the ownership of the different mobility tools, and vice versa. Persons tend to aim for compensation between the different dimensions of life, with changes concerning locations, i.e., the places of residence, education and employment, taking place significantly more frequently than changes concerning the ownership of the various mobility tools. At the same time, however, events occur increasingly simultaneously. As spatial changes take place, the actual travel behaviour seems to be reconsidered and altered. Nevertheless, it is difficult to make clear statements about the causal connection between the various aspects of long-term and mid-term mobility behaviour as well as about the influence of other dimensions of life, such as personal and familial events. In spite of the chronological order of two events the impact can be in the opposite direction, as events are anticipated in advance. So, for instance, a family moves before the birth of a child, but still due to this fact. In this case it is necessary to directly ask the respondents about the reasons for changing their place of residence.

Overall, the most important influencing variables concerning the long-term and mid-term mobility decisions in the analyses are age, gender, occupation and income as well as the personal and familial situation.

In general, persons between the ages of 15 and 35 years are most mobile, i.e., moving and changing occupation as well as varying the ownership of mobility tools most frequently. Afterwards, they become relatively established.

During the life course car ownership is highest among those who are 35 to 55 years old today. Concerning the ownership of national and regional season tickets, the opposite trend is visible. This means that car ownership on the one side and national and regional season ticket ownership on the other side substitute one another. The ownership of half-fare discount tickets increases relatively strongly over the life course.

Based on the high car ownership of the persons aged between 35 and 55 years and its stability over time, further substantial increases in the ownership of cars are to be expected. Unlike the older generations of today, which still have a relatively low level of motorisation now, the older generations of tomorrow will continue owning their cars, as they age. Since mobility tool ownership can be used as a proxy for the actual travel behaviour, there will also be a growth in overall car usage.

In order to change this development, policy and planning instruments should aim at the younger generations, as their travel habits and routines are not fully established yet, confirming findings by Prillwitz and Lanzendorf (2006), and, therefore, easier to influence. One objective should be to prevent them from buying their first car, because once a person has decided on owning and using a car, he or she is bound to this decision, by the very cost of the initial investment. Furthermore, the commitment to cars should be weakened, by providing more flexible alternatives, such as car sharing. On the other side, the commitment to public transport should be strengthened, for instance, by strategies increasing the attachment towards this mode and improving the identification, e.g., by establishing a closer and a more direct contact between the operators and the customers. Corresponding instruments, in order to encourage a shift from the motorised private transport towards more sustainable means of transport, like public transport, cycling and walking, include the provision of information about these alternatives, especially targeting younger people, implementing more successful public transport marketing strategies and temporary free or discounted public transport tickets, free bicycles, etc. At the same time, it is important to make the alternative modes of transport more attractive as well as to provide a more cyclable and walkable environment. In this context, spatial and transport planning play a crucial role. On the other side, there exist a number of strategies which aim to discourage car use. Such measures include access restrictions, parking management, road pricing and increased fuel costs (Karlaftis and Golias, 2002). Economic policy instruments are probably most effective, but least popular as well (Lanzendorf, 2006). Unfortunately, cost related information is not included in the analyses, because it was not spatially and temporally disaggregated available for the observed time period from 1985 to 2004.

Furthermore, male respondents show a more stable behaviour than female respondents. Women seem to be more flexible, for instance, making considerable adjustments following the birth of children. Once children are born and live in a household, they have a stabilising influence on the long-term and mid-term mobility. This also applies to the household size and the accommodation size. In this context, it is central to directly support the public transport usage of new mothers, since their behaviour strongly influences the socialisation of their children, as they grow up. Comfort, safety and security are important related issues.

Additional opportunities to significantly influence travel behaviour are provided by the occurrence of so-called key or life events, such as important personal and familial events as well as changes in residence, education and employment, as habits and routines are broken or at least weakened, and individuals reconsider their behaviour and consciously reflect their decisions. There exist only short periods of time during an individual's life course in which he or she looks into travel choices (Gorr, 1997).

The analyses of the long-term and mid-term mobility decisions during the life course show that these events play an important role with respect to the ownership of the various mobility tools. In this context, residential relocations as well as changes in occupation seem to be the most important ones. Therefore, these spatial alterations provide interesting starting points for policies and other interventions aiming at travel behaviour change, due to accessibility and transport systems changes (Bamberg, 2006). Bamberg (2006) carried out a study, where people received a free ticket and personal schedule information shortly after a residential relocation. He showed that, as expected, public transport use strongly increased and car use significantly decreased after the move. The same development was found by Rölle, Weber and Bamberg (2002) after distributing public transport incentives to movers. So, obviously, there exists a noteworthy potential in the population for mode changes in favour of public transport, although it is often stated that it is not possible to reverse increasing car use, because the car provides so many options and such a high degree of individual freedom (Scheiner, 2006). Interventions associated with residential relocations may be more effective in reaching higher levels of (positive) behaviour change (Stanbridge and Lyons, 2006). However, crucial to the success of any such interventions is their timing.

Another possibility related to moving is to provide information on potential locations and facilities for activities, such as shopping and leisure, in the household's neighbourhood. But this requires that such local activity places exist in the vicinity of the new place of residence, which also concerns land-use planning.

Changes in occupation, such as entering education or employment as well as retirement, offer a further opportunity to alter travel behaviour, for instance, by providing information at the place of occupation about alternative ways to travel there as well as mobility management and car pooling schemes.

Overall, it is important to notice that the explanatory variables considered in the analyses of the long-term and mid-term mobility decisions do not include those relevant for policy and planning instruments, such as costs. In addition, the ones considered are relatively difficult to influence.

10 Outlook

Further possible analyses consider the life course as a whole where the life course is seen as a sequence of various events and states (Sackmann and Wingens, 2001). The aim is to find sequence resemblances and common sequential patterns among life courses (Joh, Arentze, Hofman and Timmermans, 2002; Schlich, 2004). In this context, life courses are compared with one another and the distance between them is measured by determining the minimal number of necessary operations, such as substituting, deleting and inserting, to transform one into the other (Erzberger, 2001). This distance is then used to determine similarities and differences between life courses and to identify different life course types (Sackmann and Wingens, 2001). It is also possible to specify pre-defined or "normative" life courses and compare the observed life courses with those, measuring the deviations (Elder, 2000). Thereby, even different dimensions can be taken into account (Erzberger, 2001).

So far, the analyses concentrate on the individual. A further step is the consideration of the entire household, since lives are linked and lived interdependently among members of a family and kin (Elder, 2000). However, it is difficult to follow households over longer period of time, as they emerge and dissolve continuously (Clark *et al.*, 2003). This also applies to the tracing of individuals within households, as an individual does not a priori belong to a certain household, but rather leaves and enters households, founds new households, etc. (Scheiner, 2006). The household plays a central role, because it implies complex compromises of the individuals involved (Kaufmann, 2002). The connections between linked lives extend across people's life courses as well as across generations (Elder, 2000). In this context, Courgeau (1985) points to the "inheritability" of mobility behaviour, which probably also applies to the ownership of mobility tools.

A third temporal dimension, in addition to age and cohort membership, includes the historical context, referring to events which are independent from the individuals (Armoogum *et al.*, 2002; Mayer and Huinink, 1990). Therefore, it is interesting to consider the general situation and conditions, incorporating their development over time into the analyses of people's life courses.

Besides the exploration of the temporal dimensions, the spatial structures can also be considered. This refers to the activity spaces of individuals, their size and shape, as well as their development over time and during the life course, taking into account the different places of residence and occupation as well as the distances between those locations (Schönfelder, 2006).

Concerning the methodology, further developments combine cross-sectional and longitudinal analyses. In duration modelling these include the estimation of more flexible hazard models with the form of discrete choice models that allow for intra-individual and inter-individual variability of people (Bhat, 2003; Bhat, Srinivasan and Axhausen, 2005). These analyses can be applied for the residential mobility as well as for the ownership of mobility tools taking into account the personal and family history as well as changes in education and employment.

A further research question concerns developments in car ownership, with respect to the types of vehicles and their size, over time and over the life course. There is an increasing diversity in the car fleet held by households, shifting from small passenger cars to large non-passenger cars (Sen and Bhat, 2006). In this context, the concrete transactions are as well of interest (De Jong, 1996).

11 References

Allison, P. D. (1995) *Survival Analysis Using the SAS System: A Practical Guide*, SAS Institute Inc., Cary.

Armoogum, J., J.-L. Madre and Y. Bussière (2002) Uncertainty in long term forecasting of travel demand from demographic modelling, *Proceeding of the 13th Mini EURO Conference*, Bari, June 2002.

Ascoli, L. (2000) The index of purchasing power of the Euro, *Statistics in focus: Theme 2*, **37**, 1-8.

Aufhauser, E. (1995) Wohnchancen – Wohnrisiken: Räumliche Mobilität und wohnungsbezogene Lebensführung in Wien im gesellschaftlichen Wandel, *Abhandlungen zur Geographie und Regionalforschung*, **4**, Institut für Geographie der Universität Wien, Wien.

Axhausen, K. W. (2007) Predicting response rate: A natural experiment, *Arbeitsberichte Verkehrs- und Raumplanung*, **434**, Institut für Verkehrsplanung und Transportsysteme (IVT), ETH Zürich, Zürich.

Axhausen, K. W. and S. Beige (2003) Besitz von Mobilitätsressourcen und deren Nutzung sowie Änderungen des Wohnortes, Forschungsprogramm UNIVOX 2003 Teil I G Verkehr, Trendbericht, GfS-Forschungsinstitut, Zürich.

Axhausen, K. W., A. Simma and T. Golob (2001) Pre-commitment and usage: Cars, season-tickets and travel, *European Research in Regional Science*, **11**, 101-110.

Bamberg, S. (2006) Is a residential relocation a good opportunity to change people's travel behavior? Results from a theory-driven intervention study, *Environment and Behavior*, **38** (6) 820-840.

Beck, U. (1986) *Risikogesellschaft: Auf dem Weg in eine andere Moderne*, Suhrkamp, Frankfurt/Main.

Beck, U. and E. Beck-Gernsheim (2002) *Individualization: Institutionalized Individualism and Its Social and Political Consequences*, Sage Publications, London.

Beige, S. (2004) Ownership of mobility tools in Switzerland, presentation at the *4th Swiss Transport Research Conference*, Ascona, March 2004.

Beige, S. (2006) Long-term and mid-term mobility decisions during the life course, presentation at the *6th Swiss Transport Research Conference*, Ascona, March 2006.

Beige, S. and K. W. Axhausen (2005) Feldbericht der Befragung zur langfristigen räumlichen Mobilität, *Arbeitsberichte Verkehrs- und Raumplanung*, **315**, Institut für Verkehrsplanung und Transportsysteme (IVT), ETH Zürich, Zürich.

Bekhor, S. (1999) Integration of behavioral transportation planning: Models with the traffic assignment problem, PhD thesis, Technion, Haifa.

Ben-Akiva, M. and M. Bierlaire (1999) Discrete choice methods and their application to short term travel decisions, in R. Hall (ed.) *Handbook of Transportation Science*, 5-34, Kluwer Academic Publishers, Boston.

Ben-Akiva, M. and S. R. Lerman (1985) *Discrete Choice Analysis: Theory and Application to Travel Demand*, MIT Press, Cambridge.

Bhat, C. R. (2003) Econometric choice formulations: Alternative model structures, estimation techniques and emerging directions, paper presented at the *10th International Conference on Travel Behaviour Research*, Lucerne, August 2003.

Bhat, C. R. and S. Sen (2006) Household vehicle type holdings and usage: An application of the multiple discrete-continuous extreme value (MDCEV) model, *Transportation Research Part B*, **40** (1) 35-53.

Bhat, C. R., S. Srinivasan and K. W. Axhausen (2005) An analysis of multiple interactivity durations using a unifying multivariate hazard model, *Transportation Research Part B*, **39** (9) 797-824.

Bierlaire, M. (2001) A general formulation of the cross-nested logit model, presentation at the *1st Swiss Transport Research Conference*, Ascona, March 2001.

Bierlaire, M. (2005) *An introduction to BIOGEME (Version 1.4)*, Institute of Urban and Regional Planning and Design, EPFL, Lausanne.

Bird, K., C. Born and C. Erzberger (2000) Ein Bild des eigenen Lebens zeichnen: Zum Einsatz eines Kalenders als Visualisierungsinstrument zur Erfassung individueller Lebensverläufe, *Sfb-Arbeitspapier*, **59**, Universität Bremen, Bremen.

Birg, H. and E. J. Flöthmann (1992) Biographische Determinanten der räumlichen Mobilität, in Akademie für Raumforschung und Landesplanung (ed.) Regionale und biographische Mobilität im Lebensverlauf, *Forschungs- und Sitzungsberichte*, **189**, 27-52, Akademie für Raumforschung und Landesplanung, Hannover.

Bjørner, T. B. and S. Leth-Petersen (2004) The effect on car ownership of changes in household size and location: Descriptive analyses based on panel household data, *AKF Working Paper*, Danish Institute of Governmental Research, Copenhagen.

Bjørner, T. B. and S. Leth-Petersen (2005) Dynamic models of car ownership at the household level, *International Journal of Transport Economics*, **23** (1) 57-75.

Blijie, B. (2005) The impact of accessibility on residential choice: Empirical results of a discrete choice model, paper presented at the *45th European Regional Science Association*, Amsterdam, August 2005.

Blossfeld, H.-P. and J. Huinink (2001) Lebensverlaufsforschung als sozialwissenschaftliche Forschungsperspektive: Themen, Konzepte, Methoden und Probleme, *BIOS. Zeitschrift für Biographieforschung und Oral History*, **14** (2) 5-31.

Box-Steffensmeier, J. M. and B. S. Jones (2004) *Event History Modeling: A Guide for Social Scientists*, Cambridge University Press, Cambridge.

Brandtstädter, J. (1990) Entwicklung im Lebensverlauf: Ansätze und Probleme der Lebensspannen – Entwicklungspsychologie, in K. U. Mayer (ed.) Lebensverläufe und sozialer Wandel, *Kölner Zeitschrift für Soziologie und Sozialpsychologie*, Sonderheft **31**, 322-350.

Brückner, E. (1990) Die retrospektive Erhebung von Lebensverläufen, in K. U. Mayer (ed.) Lebensverläufe und sozialer Wandel, *Kölner Zeitschrift für Soziologie und Sozialpsychologie*, Sonderheft **31**, 374-403.

Brückner, E. (1994) Erhebung ereignisorientierter Lebensverläufe als retrospektive Längsschnittrekonstruktion, in R. Hauser, N. Ott and G. Wagner (eds.) Erhebungsverfahren, Analysemethoden und Mikrosimulation, *Mikroanalytische Grundlagen der Gesellschaftspolitik*, **2**, 38-69, Akademie-Verlag, Berlin.

Bussière, Y., J. Armoogum, C. Gallez, C. Girard and J.-L. Madre (1994) Longitudinal approach to motorization: Long term dynamics in three urban regions, paper presented at the *7th International Conference on Travel Behaviour Research*, Valle Nevada, June 1994.

Cao, X., P. L. Mokhtarian and S. L. Handy (2006) Examining the impacts of residential self-selection on travel behavior: Methodologies and empirical findings, *Research Report*, **UCD-ITS-RR-06-18**, Institute of Transportation Studies, University of California, Davis.

Carle, G. (2005) Erdgasfahrzeuge im Wettbewerb, *Arbeitsberichte Verkehrs- und Raumplanung*, **269**, Institut für Verkehrsplanung und Transportsysteme (IVT), ETH Zürich, Zürich.

Cascetta, E. (2001) *Transportation Systems Engineering: Theory and Methods*, Kluwer Academic Publishers, Dordrecht.

Chapin, F. S., Jr. (1965) *Urban Land Use Planning*, University of Illinois, Urbana.

Chapin, F. S., Jr. (1974) *Human Activity Patterns in the City*, John Wiley & Sons, New York.

Clark, W. A. V. and J. L. Onaka (1983) Life cycle and housing adjustment as explanations of residential mobility, *Urban Studies*, **20** (1) 47-57.

Clark, W. A. V., M. C. Deurloo and F. M. Dieleman (2003) Housing careers in the United States, 1968-93: Modelling the sequencing of housing states, *Urban Studies*, **40** (1) 143-160.

Courgeau, D. (1985) Interaction between spatial mobility, family and career life-cycle: A French survey, *European Sociological Review*, **1** (2) 139-162.

Cox, D. R. (1972) Regression models and life tables, *Journal of the Royal Statistical Society Series B*, **34** (2) 187-220.

Cox, D. R. (1975) Partial likelihood, *Biometrika*, **62** (2) 269-276.

Dargay, J. M. (2001) The effect of income on car ownership: Evidence of asymmetry, *Transportation Research Part A*, **35** (9) 807-821.

Dargay, J., L. Hivert and D. Legros (2006) An investigation of car ownership in Europe based on the European Community Household Panel, paper presented at the *11th International Conference on Travel Behaviour Research*, Kyoto, August 2006.

De Jong, G. (1996) A disaggregate model system of vehicle holding duration, type choice and use, *Transportation Research Part B*, **30** (4) 263-276.

De Jong, G., J. Fox, A. Daly, M. Pieters and R. Smit (2004) Comparison of car ownership models, *Transport Review*, **24** (4) 379-408.

Dex, S. (1991) The reliability of recall data: A literature review, *Working papers*, **11**, ESRC Research Centre on Micro-Social Change, Colchester.

Diekmann, A. (1995) *Empirische Sozialforschung: Grundlagen, Methoden, Anwendungen*, Rowohlt Taschenbuch Verlag, Reinbek bei Hamburg.

Dieleman, F. M. (2001) Modelling residential mobility: A review of recent trends in research, *Journal of Housing and the Built Environment*, **16** (3-4) 249-265.

Dillman, D. A. (2000) *Mail and Internet Surveys: The Tailored Design Method*, John Wiley & Sons, New York.

Elder, G. H., Jr. (1995) The life course paradigm: Social change and individual development, in P. Moen, G. H. Elder, Jr., and K. Lüscher (eds.) *Examining Lives in Context: Perspectives on the Ecology of Human Development*, 101-139, American Psychological Association, Washington, DC.

Elder, G. H., Jr. (2000) The life course, in E. F. Borgatta (ed.) *Encyclopedia of Sociology*, 1614-1622, Macmillan Reference, New York.

Erzberger, C. (2001) Sequenzmusteranalyse als fallorientierte Analysestrategie, in R. Sackmann und M. Wingens (ed.) Strukturen des Lebenslaufs: Übergang, Sequenz, Verlauf, *Statuspassagen und Lebenslauf*, **1**, 135-162, Juventa Verlag, Weinheim.

Federal Office for Spatial Development (2002) Verkehrliche Raumgliederung (V1-V5) ausgehend von der "Raumgliederung 2002" (1-13), Bundesamt für Raumentwicklung, Bern.

Federal Statistical Office (2000) Census of the year 2000, Bundesamt für Statistik, Neuchâtel.

Franz, P. (1984) *Soziologie der räumlichen Mobilität: Eine Einführung*, Campus Verlag, Frankfurt/Main.

Freedman, D., A. Thornton, D. Camburn, D. Alwin and L. Young-DeMarco (1988) The life history calendar: A technique for collecting retrospective data, *Sociological Methodology*, **18**, 37-68.

Gärling, T. and K. W. Axhausen (2003) Introduction: Habitual travel choice. *Transportation*, **30** (1) 1-11.

Gorr, H. (1997) *Die Logik der individuellen Verkehrsmittelwahl: Theorie und Realität des Entscheidungsverhaltens im Personenverkehr*, Focus Verlag, Giessen.

Hägerstrand, T. (1970) What about people in regional science?, *Papers in Regional Science*, **24** (1) 7-24.

Han, A. and J. A. Hausman (1990) Flexible parametric estimation of duration and competing risk models, *Journal of Applied Econometrics*, **5** (1) 1-28.

Heckhausen, J. (1990) Erwerb und Funktion normativer Vorstellungen über den Lebenslauf: Ein entwicklungspsychologischer Beitrag zur sozio-psychischen Konstruktion von Biographien, in K. U. Mayer (ed.) Lebensverläufe und sozialer Wandel, *Kölner Zeitschrift für Soziologie und Sozialpsychologie*, Sonderheft **31**, 351-373.

Heine, H., R. Mautz and W. Rosenbaum (2001) *Mobilität im Alltag: Warum wir nicht vom Auto lassen*, Campus Verlag, Frankfurt/Main.

Hensher, D. A. (1998) The timing of change for automobile transactions: Competing risk multispell specification, in J. de Dios Ortúzar, D. A. Hensher and S. Jara-Diaz (eds.) *Travel Behaviour Research: Updating the State of Play*, 487-506, Elsevier, Oxford.

Hensher, D. A., J. M. Rose and W. H. Greene (2005) *Applied Choice Analysis: A Primer*, Cambridge University Press, Cambridge.

Hollingworth, B. J. and E. J. Miller (1996) Retrospective interviewing and its application in study of residential mobility, *Transportation Research Record*, **1551**, 74-81.

Hsiao, C. (2003) *Analysis of Panel Data*, Cambridge University Press, Cambridge.

Joh, C.-H., T. Arentze, F. Hofman and H. Timmermans (2002) Activity pattern similarity: A multidimensional sequence alignment method, *Transportation Research Part B*, **36** (5) 385-403.

Jones, P. M., M. C. Dix, M. I. Clarke and I. G. Heggie (1983) *Understanding Travel Behaviour*, Gower Publishing, Aldershot.

Kalbfleisch, J. D. and R. L. Prentice (1980) *The Statistical Analysis of Failure Time Data*, John Wiley & Sons, New York.

Karlaftis, M. and J. Golias (2002) Automobile ownership, households without automobiles, and urban traffic parameters: Are they related?, *Transportation Research Record*, **1792**, 29-35.

Kaufmann, V. (2002) *Re-thinking Mobility: Contemporary Sociology*, Ashgate, Aldershot.

Kelle, U. and S. Kluge (2001) Einleitung, in S. Kluge and U. Kelle (eds.) Methodeninnovation in der Lebenslaufforschung: Integration qualitativer und quantitativer Verfahren in der Lebenslauf- und Biographieforschung, *Statuspassagen und Lebenslauf*, **4**, 11-33, Juventa Verlag, Weinheim.

Kendig, H. L. (1984) Housing careers, life cycle and residential mobility: Implications for the housing market, *Urban Studies*, **21** (3) 271-283.

Keupp, H. and B. Röhrle (1987) *Soziale Netwerke*, Campus Verlag, Frankfurt/Main.

Klein, T. and D. Fischer-Kerli (2000) Die Zuverlässigkeit retrospektiv erhobener Lebensverlaufsdaten: Analysen zur Partnerschaftsbiographie des Familiensurvey, *Zeitschrift für Soziologie*, **29** (4) 294-312.

Ladkin, A. (2002) Autobiographical memory and life history data: Te relevance of this approach for tourism and hospitality research, paper presented at the *British-German Conference on Tourism Research*, Münster, September 2002.

Lanzendorf, M. (2003) Mobility biographies: A new perspective for understanding travel behaviour, paper presented at the *10th International Conference on Travel Behaviour Research*, Lucerne, August 2003.

Lanzendorf, M. (2004) Stability and instability of mobility biographies: An explorative analysis, presentation at the *EIRASS Workshop*, Maastricht, May 2004.

Lanzendorf, M. (2006) Key events and their effect on mobility biographies: The case of child birth, paper presented at the *11th International Conference on Travel Behaviour Research*, Kyoto, August 2006.

Lelièvre, E. and C. Bonvalet (1994) A compared cohort history of residential mobility, social change and home-ownership in Paris and the rest of France, *Urban Studies*, **31** (10) 1647-1665.

Löchl, M., M. Bürgle and K. W. Axhausen (2007) Implementierung des integrierten Flächennutzungsmodells UrbanSim für den Grossraum Zürich: Ein Erfahrungsbericht, *DISP*, **168**, 13-25.

Lück, D., R. Limmer and W. Bonß (2006) Theoretical approaches to job mobility, in E. Widmer and N. F. Schneider (eds.) State-of-the-Art of Mobility Research: A Literature Analysis for Eight Countries, *JobMob and FamLives Working Paper*, **No. 2006-01**, 5-42, available on http://www.jobmob-and-famlives.eu.

Maier, G. and P. Weiss (1990) *Modelle diskreter Entscheidungen: Theorie und Anwendung in den Sozial- und Wirtschaftswissenschaften*, Springer-Verlag, Wien.

Mayer, K. U. (1990) Lebensverläufe und sozialer Wandel: Anmerkungen zu einem Forschungsprogramm, in K. U. Mayer (ed.) Lebensverläufe und sozialer Wandel, *Kölner Zeitschrift für Soziologie und Sozialpsychologie*, Sonderheft **31**, 7-21.

Mayer, K. U. and J. Huinink (1990) Alters-, Perioden- und Kohorteneffekte in der Analyse von Lebensverläufen Oder: Lexis Ade?, in K. U. Mayer (ed.) Lebensverläufe und sozialer Wandel, *Kölner Zeitschrift für Soziologie und Sozialpsychologie*, Sonderheft **31**, 442-459.

McFadden, D. (1978). Modelling the choice of residential location, in A. Karlquist, L. Lundquist, F. Snickars and J. J.Weibull (eds.) *Spatial Interaction Theory and Residential Location*, 75-96, North Holland, Amsterdam.

Morrow-Jones, H. A. and M. V. Wenning (2005) The housing ladder, the housing life-cycle and the housing life-course: Upward and downward movement among repeat home-buyers in a US metropolitan housing market, *Urban Studies*, **42** (10) 1739-1754.

Ohnmacht, T. (2006) Die Geografie des Sozialen als Aktivitätsraum: Räumliche Verteilung der Sozialkontakte unter den Bedingungen von Mobilitätsbiografien, Diplomarbeit, Institut für Verkehrsplanung und Transportsysteme (IVT), ETH Zürich, Zürich.

Olsson, C. (2005) The index of purchasing power of the Euro, *Statistics in focus: Economy and Finance*, **23**, 1-8.

Ortúzar, J. de D. and L. G. Willumsen (2001) *Modelling Transport*, John Wiley & Sons, Chichester.

Peters, H. E. (1988) Retrospective versus panel data in analyzing lifecycle events, *Journal of Human Resources*, **23** (4) 488-513.

Pooley, C. G., J. Turnbull and M. Adams (2005) *A Mobile Century? Changes in Everyday Mobility in Britain in the Twentieth Century*, Ashgate, Aldershot.

Prillwitz, J., S. Harms and M. Lanzendorf (2006) Impact of life course events on car ownership, *Transportation Research Record*, **1985**, 71-77.

Prillwitz, J. and M. Lanzendorf (2006) The importance of life course events for daily travel behaviour: A panel analysis, paper presented at the *11th International Conference on Travel Behaviour Research*, Kyoto, August 2006.

Ravenstein, E. G. (1885) The laws of migration, *Journal of the Statistical Society*, **48** (2) 167-235.

Ravenstein, E. G. (1889) The laws of migration, *Journal of the Royal Statistical Society*, **52** (2) 241-305.

Rölle, D., C. Weber and S. Bamberg (2002) Mögliche Beiträge von Verkehrsverminderung und -verlagerung zu einem umweltgerechten Verkehr in Baden-Württemberg – eine empirische Analyse der Bestimmungsfaktoren von Haushaltsentscheidungen, *Forschungsbericht*, Institut für Rationelle Energiewirtschaft und Energieanwendung, Universität Stuttgart, Stuttgart.

Rossi, P. H. (1955) *Why Families Move*, The Free Press, Glencoe.

Rouwendal, J. and A. van der Vlist (2005) A dynamic model of commutes, *Environment and Planning A*, **37** (12) 2209-2232.

Ryder, N. B. (1965) The cohort as a concept in the study of social change, *American Sociological Review*, **30** (6) 843-861.

Sackmann, R. and M. Wingens (2001) Theoretische Konzepte des Lebenslaufs: Übergang, Sequenz und Verlauf, in R. Sackmann and M. Wingens (eds.) Strukturen des Lebenslaufs: Übergang, Sequenz, *Verlauf, Statuspassagen und Lebenslauf*, **1**, 17-48, Juventa Verlag, Weinheim.

Salomon, I. and M. Ben-Akiva (1983) The use of the life-style concept in travel demand models, *Environment and Planning A*, **15** (5) 623-638.

Scheiner, J. (2006) Housing mobility and travel behaviour: A process-oriented approach to spatial mobility: Evidence from a new research field in Germany, *Journal of Transport Geography*, **14** (4) 287-298.

Schlich, R. (2004) Verhaltenshomogene Gruppen in Längsschnitterhebungen, Dissertation, Institut für Verkehrsplanung und Transportsysteme (IVT), ETH Zürich, Zürich.

Schönfelder, S. (2006) Urban rhythms: Modelling the rhythms of individual travel behaviour, Dissertation, Institut für Verkehrsplanung und Transportsysteme (IVT), ETH Zürich, Zürich.

Scott, J. (2000) *Social Network Analysis*, Sage Publications, London.

Scott, J. and D. Alwin (1998) Retrospective versus prospective measurement of life histories in longitudinal research, in J. Z. Giele and G. H. Elder, Jr. (eds.) *Methods of Life Course Research: Qualitative and Quantitative Approaches*, 98-127, Sage Publications, Thousand Oaks.

Scott, D. M. and K. W. Axhausen (2006) Household mobility tool ownership: Modeling interactions between cars and season tickets, *Transportation*, **33** (4) 311-328.

Sen, S. and C. R. Bhat (2006) A comprehensive analysis of household vehicle make/model/vintage and usage decisions, paper presented at the *11th International Conference on Travel Behaviour Research*, Kyoto, August 2006.

Simma, A. and K. W. Axhausen (2001) Structures of commitment in mode use: A comparison of Switzerland, Germany and Great Britain, *Transport Policy*, **8** (4) 279-288.

Simma, A. and K. W. Axhausen (2003) Commitments and modal usage: Analysis of German and Dutch panels, *Transportation Research Record*, **1854**, 22-31.

Spiegel, E. (1992) Biographische und räumliche Mobilität jüngerer Erwachsenenhaushalte in innenstadtnahen Wohngebieten: Fallstudie Hamburg, in Akademie für Raumforschung und Landesplanung (ed.) Regionale und biographische Mobilität im Lebensverlauf, *Forschungs- und Sitzungsberichte*, **189**, 125-148, Akademie für Raumforschung und Landesplanung, Hannover.

Stanbridge, K. and G. Lyons (2006) Travel behaviour considerations during the process of residential relocation, paper presented at the *11th International Conference on Travel Behaviour Research*, Kyoto, August 2006.

Statistical Office of the European Communities (2006) Europa in Zahlen: Eurostat Jahrbuch, http://epp.eurostat.ec.europa.eu/portal/page?_pageid=1090,30070682,1090_33076576 &_dad=portal&_schema=PORTAL, April 2006.

Swiss Household Panel (2005) "Living in Switzerland 1999-2020" project, Swiss National Science Foundation, Federal Statistical Office and University of Neuchâtel, Neuchâtel.

Tuma, N. B. and M. T. Hannan (1984) *Social Dynamics: Models and Methods*, Academic Press, Orlando.

Vandersmissen, M. H., A. M. Séguin, M. Thériault and C. Claramunt (2005) Modelling propensity to move house after job change using event history analysis and GIS, paper presented at the *2nd MCRI/GEOIDE PROCESSUS Colloquium*, Toronto, June 2005.

Van der Waerden, P., H. Timmermans and A. Borgers (2003) The influence of key events and critical incidents on transport mode choice switching behaviour: A descriptive analysis, paper presented at the *10th International Conference on Travel Behaviour Research*, Lucerne, August 2003.

Van Ommeren, J., P. Rietveld and P. Nijkamp (1999) Job moving, residential moving, and commuting: A search perspective, *Journal of Urban Economics*, **46** (2) 230-253.

Van Ommeren, J., P. Rietveld and P. Nijkamp (2000) Job mobility, residential mobility and commuting: A theoretical analysis using search theory, *The Annals of Regional Science*, **34** (2) 213-232.

Verhoeven, M., T. Arentze, H. Timmermans and P. van der Waerden (2005) Modeling the impact of key events on long-term transport mode choice decisions: Decision network approach using event history data, *Transportation Research Record*, **1926**, 106-114.

Vrtic, M. (2003) Simultanes Routen- und Verkehrsmittelwahlmodell, Dissertation, Fakultät für Verkehrswissenschaften, TU Dresden, Dresden.

Vrtic, M., P. Fröhlich, N. Schüssler, K. W. Axhausen, S. Dasen, S. Erne, B. Singer, D. Lohse and C. Schiller (2005) Erzeugung neuer Quell-/Zielmatrizen im Personenverkehr, report to the Bundesamt für Raumentwicklung and the Bundesamt für Strassen und für Verkehr, Institut für Verkehrsplanung und Transportsysteme (IVT), ETH Zürich, Zürich, Emch + Berger AG, Zürich, and TU Dresden, Dresden.

Vrtic, M., P. Fröhlich, N. Schüssler, K. W. Axhausen, C. Schulze, P. Kern, F. Perret, S. Pfisterer, C. Schultze, A. Zimmerman and U. Heidl (2005) Verkehrsmodell für den öffentlichen Verkehr des Kantons Zürich, report to the Amt für Verkehr des Kantons Zürich, Institut für Verkehrsplanung und Transportsysteme (IVT), ETH Zürich, Zürich, Ernst Basler + Partner AG, Zürich, and PTV AG, Karlsruhe.

Wagner, M. (1990) Wanderungen im Lebensverlauf, in K. U. Mayer (ed.) Lebensverläufe und sozialer Wandel, *Kölner Zeitschrift für Soziologie und Sozialpsychologie*, Sonderheft **31**, 212-238.

Wagner, M. (1992) Zur Bedeutung räumlicher Mobilität für den Erwerbsverlauf bei Männern und Frauen, in Akademie für Raumforschung und Landesplanung (ed.) Regionale und biographische Mobilität im Lebensverlauf, *Forschungs- und Sitzungsberichte*, **189**, 149-167, Akademie für Raumforschung und Landesplanung, Hannover.

Waldner, U., M. Löchl and M. Bürgle (2005) Haushaltsbefragung zur Wohnsituation im Grossraum Zürich – Feldbericht, *Arbeitsberichte Polyprojekt Zukunft urbane Kulturlandschaften*, **1**, Netzwerk Stadt und Landschaft (NSL), ETH Zürich, Zürich.

Weißhuhn, G. and F. Büchel (1992) Betriebswechsel, räumliche Mobilität und Verdienstentwicklung: Eine Longitudinalanalyse sozialversicherungspflichtig Beschäftigter in den Perioden 1977-1979 und 1982-1984, in Akademie für Raumforschung und Landesplanung (ed.) Regionale und biographische Mobilität im Lebensverlauf, *Forschungs- und Sitzungsberichte*, **189**, 168-199, Akademie für Raumforschung und Landesplanung, Hannover.

Yamaguchi, K. (1991) *Event History Analysis*, Sage Publications, Newbury Park.

Zondag, B. and M. Pieters (2005) Influence of accessibility on residential location choice, *Transportation Research Record*, **1902**, 63-70.

Zumkeller, D., J.-L. Madre, B. Chlond and J. Armoogum (2006) Panel surveys, in P. Stopher and C. Stecher (eds.) *Travel Survey Methods: Quality and Future Directions*, 363-398, Elsevier, Oxford.

Appendix

Appendix A: Questionnaire of the retrospective survey (household form and person form)

A 1 Questionnaire in German (original version)

A 2 Questionnaire in English

Appendix B: Models for the mobility tool ownership

B 1 Models for the mobility tool ownership for the year 2005

B 2 Models for the mobility tool ownership for the period from 1985 to 2004

Teil 1: Haushaltsfragebogen

Bitte geben Sie die Adresse Ihres Wohnortes an.

Strasse und Hausnummer	
Postleitzahl	Gemeinde

Bitte geben Sie für jede Person in Ihrem Haushalt Geburtsjahr, Geschlecht sowie den Ort des gegenwärtigen Ausbildungs- bzw. Arbeitsplatzes an und ob es sich dabei um eine Ausbildung oder eine Arbeit handelt.

Geburtsjahr	Geschlecht männlich	weiblich	Gegenwärtiger Ausbildungs- bzw. Arbeitsplatz PLZ	Gemeinde	
	☐	☐			☐ Ausbildung ☐ Arbeit
	☐	☐			☐ Ausbildung ☐ Arbeit
	☐	☐			☐ Ausbildung ☐ Arbeit
	☐	☐			☐ Ausbildung ☐ Arbeit
	☐	☐			☐ Ausbildung ☐ Arbeit
	☐	☐			☐ Ausbildung ☐ Arbeit
	☐	☐			☐ Ausbildung ☐ Arbeit
	☐	☐			☐ Ausbildung ☐ Arbeit
	☐	☐			☐ Ausbildung ☐ Arbeit

Wie hoch ist das gesamte monatliche Brutto-Einkommen Ihres Haushaltes?

- ☐ unter 2 000 Fr.
- ☐ 2 000 bis 3 999 Fr.
- ☐ 4 000 bis 5 999 Fr.
- ☐ 6 000 bis 7 999 Fr.
- ☐ 8 000 bis 9 999 Fr.
- ☐ 10 000 bis 11 999 Fr.
- ☐ 12 000 bis 13 999 Fr.
- ☐ 14 000 Fr. und mehr

Wie viele Fahrzeuge gibt es in Ihrem Haushalt?

	Autos	☐ Keine Autos		Motorräder ab 125 ccm	☐ Keine Motorräder
	betriebsbereite Velos	☐ Keine Velos		Kleinmotorräder bis 125 ccm	☐ Keine Kleinmotorräder

Teil 2: Personenfragebogen

Eidgenössische Technische Hochschule Zürich
Swiss Federal Institute of Technology Zurich

Wann sind Sie geboren?

Monat [ ]   Jahr [ ]

Sind Sie ...

☐ männlich   ☐ weiblich

Sind Sie ...

☐ Schweizer/Schweizerin
☐ Von anderer Nationalität, und zwar [ ]

Welche Ausbildungsabschlüsse haben Sie erworben? (Mehrfachnennungen möglich)

☐ Primarschulabschluss   ☐ Lehrabschluss   ☐ Fachhochschulabschluss
☐ Sekundar- oder Realschulabschluss   ☐ Maturitätsabschluss   ☐ Universitätsabschluss/ETH-Abschluss

Sind Sie ...

☐ Vollzeit berufstätig   ☐ in Ausbildung   ☐ Hausmann/Hausfrau
☐ Teilzeit berufstätig   ☐ auf Arbeitssuche   ☐ Rentner/Rentnerin (AHV/IV)

Welcher Ausbildung bzw. Arbeit gehen Sie hauptsächlich nach?

Ausbildung [ ]   Arbeit [ ]

Wie viele Stunden gehen Sie in der Woche Ihrer Ausbildung bzw. Arbeit nach?

Ausbildung [ ] Stunden   Arbeit [ ] Stunden

Wo genau befindet sich Ihr gegenwärtiger Ausbildungs- bzw. Arbeitsplatz?
Bitte geben Sie Postleitzahl, Gemeinde, Strasse und Hausnummer an.

Ausbildungsplatz [ ]   Arbeitsplatz [ ]

Wie lange benötigen Sie bzw. würden Sie für den Weg (von Tür zu Tür) zu Ihrem Ausbildungs- bzw. Arbeitsplatz benötigen, wenn Sie ...

	Weg zum Ausbildungsplatz	Weg zum Arbeitsplatz
ausschliesslich das Auto benutzen?	ca. [ ] Minuten	ca. [ ] Minuten
ausschliesslich öffentliche Verkehrsmittel benutzen?	ca. [ ] Minuten	ca. [ ] Minuten
Weg ist mit öffentlichen Verkehrsmitteln nicht möglich.	☐	☐

Besitzen Sie einen Auto-Führerausweis? Und wann haben Sie diesen erworben?

☐ nein   ☐ ja   Jahr des Erwerbes [ ]

Teil 2: Personenfragebogen

Steht Ihnen ein Auto zum Selbstfahren zur Verfügung?

☐ immer ☐ häufig ☐ selten ☐ nie ☐ über Mobility

Wie zufrieden sind Sie ...

	sehr zufrieden									sehr unzufrieden
	1	2	3	4	5	6	7	8	9	10
mit Ihrer Gesundheit?	☐	☐	☐	☐	☐	☐	☐	☐	☐	☐
mit Ihrer Wohnung?	☐	☐	☐	☐	☐	☐	☐	☐	☐	☐
mit Ihrer Arbeit?	☐	☐	☐	☐	☐	☐	☐	☐	☐	☐
mit Ihrer Freizeit?	☐	☐	☐	☐	☐	☐	☐	☐	☐	☐
mit dem Zustand der Umwelt in Ihrer Region?	☐	☐	☐	☐	☐	☐	☐	☐	☐	☐
mit Ihrem Leben, alles in allem?	☐	☐	☐	☐	☐	☐	☐	☐	☐	☐

Werden Sie innerhalb des nächsten Jahres umziehen?

☐ sehr wahrscheinlich ☐ wahrscheinlich ☐ unwahrscheinlich ☐ sehr unwahrscheinlich

Fragen zu Ihrer Wohngeschichte

Wir möchten Sie bitten, den folgenden Lebensverlaufskalender für den Zeitraum von 1985 bis 2004 auszufüllen. Beachten Sie dabei bitte die folgenden Hinweise zum Ausfüllen des Lebensverlaufskalenders.

- Ein **Beispiel** für einen bereits ausgefüllten Lebensverlaufskalender ist auf der folgenden Seite abgebildet.
- Danach finden Sie zwei Lebensverlaufskalender. Bitte füllen Sie **einen** davon vollständig aus. Erfahrungsgemäss verschreibt man sich beim Ausfüllen leicht. Daher haben wir ein Reserveexemplar beigelegt für den Fall, dass die entsprechenden Korrekturen zu unübersichtlich werden. Bitte streichen Sie in diesem Fall die falsch ausgefüllte Version deutlich sichtbar durch.
- Beim Ausfüllen empfiehlt es sich, mit dem Eintragen der Zeitpunkte von wichtigen Ereignissen der Familiengeschichte zu beginnen (z.B. Geburt von Geschwistern, Auszug aus dem Elternhaus, Heirat, Scheidung, Geburt von Kindern, Todesfälle in der Familie, Pensionierung, usw.). Dabei kann es hilfreich sein, zuerst die neueren Ereignisse einzutragen und dann rückwärts vorzugehen.
- Im Mittelpunkt dieser Befragung steht in erster Linie Ihre Wohngeschichte während der letzten zwanzig Jahre. Daher möchten wir Sie bitten, Ihre Wohnortswechsel während dieses Zeitraumes im Lebensverlaufskalender zu kennzeichnen, dabei die Wohnorte durchzunummerieren und dann die weiteren Fragen zu den einzelnen Wohnorten auf Seite 8 zu beantworten. Falls die dort vorgegebenen fünf Wohnorte nicht ausreichen, wählen Sie bitte die für Sie wichtigsten fünf Wohnorte aus und notieren Sie dazu die entsprechende Nummer des Wohnortes aus dem Lebensverlaufskalender.
- Die Fragen zu den Wohnorten umfassen unter anderem die Adresse des Wohnortes. Bitte geben Sie nach Möglichkeit Postleitzahl, Gemeinde, Strasse und Hausnummmer an. Falls Sie sich nicht mehr genau an die Adresse erinnern können, geben Sie bitte den Namen des Ortes an und bei grösseren Städten den Namen des Stadtteiles.
- Auch wenn Sie während des Zeitraumes von 1985 bis 2004 im Ausland gelebt haben, interessieren uns die entsprechenden Angaben.

Wann sind Sie das letzte Mal vor 1985 umgezogen?

[____] Jahr des letzten Umzuges vor 1985
☐ Ich bin vor 1985 nicht umgezogen.

Wann haben Sie das letzte Mal vor 1985 den Ausbildungs- bzw. Arbeitsplatz gewechselt?

[____] Jahr des letzten Ausbildungs- bzw. Arbeitsplatzwechsels oder Ausbildungsbeginns vor 1985
☐ Ich habe vor 1985 den Ausbildungs- bzw. Arbeitsplatz nicht gewechselt.
☐ Ich bin vor 1985 keiner Ausbildung bzw. Arbeit nachgegangen.

Teil 2: Personenfragebogen

Beispiel für einen ausgefüllten Lebensverlaufskalender

	1985	1986	1987	1988	1989	1990	1991	1992
Angaben zu Ihrer Familiengeschichte								
Eintragen der Zeitpunkte von wichtigen familiären Ereignissen (z.B. Geburt von Geschwistern, Auszug aus dem Elternhaus, Heirat, Scheidung, Geburt von Kindern, Todesfälle in der Familie, Pensionierung, usw.)				|Auszug aus dem Elternhaus	|Heirat	|Geburt des 1. Kindes |Geburt des
Anzahl der Personen in Ihrem Haushalt	├─ 4 ─┤		├─ 1 ─┤		├─ 2 ─┤		├─ 3 ─┤	├─ 4 ─
Angaben zu Ihren Wohnorten (weitere Angaben auf Seite 8)								
Nummerierung der Wohnorte	├─1. Wohnort─┤		├─2. Wohnort─┤		├─3. Wohnort─┤			├─4. Wohno
Angaben zu Ihrem Besitz von Autos und ÖV-Abonnementen								
Verfügbarkeit eines Autos: immer verfügbar						├─────┤		
Verfügbarkeit eines Autos: teilweise verfügbar	├───────┤							
Verfügbarkeit eines Autos: nie verfügbar			├───────┤					
Besitz eines Halbtaxabonnements	├───┤							
Besitz eines Generalabonnements								
Besitz eines Jahres- oder Monatsabonnements	├───────┤							
Angaben zu Ihrem Ausbildungs- bzw. Arbeitsplatz								
Postleitzahl und Gemeinde des Ausbildungsplatzes	├─ 8090 Zürich ─┤							
Postleitzahl und Gemeinde des Arbeitsplatzes			├─── 8001 Zürich ───┤			├─── 8303 Bassersdorf ───		
Meistbenutztes Verkehrsmittel auf dem Weg zum Ausbildungs- bzw. Arbeitsplatz								
Auto, Motorrad, Moped, Mofa						├─────┤		
Eisenbahn, Tram, Bus	├───────┤							
Velo			├───┤					
zu Fuss								
Persönliches monatliches Brutto-Einkommen (Fremdwährung gegebenenfalls umrechnen)								
unter 2 000 Fr.	├───────┤							
2 000 bis 5 999 Fr.			├───────┤					
6 000 bis 9 999 Fr.						├─────┤		
10 000 bis 13 999 Fr.								
14 000 Fr. und mehr								

Lebensverlaufskalender
Beachten Sie bitte die Hinweise auf Seite 2.

	1985	1986	1987	1988	1989	1990	1991
Angaben zu Ihrer Familiengeschichte							
Eintragen der Zeitpunkte von wichtigen familiären Ereignissen (z. B. Geburt von Geschwistern, Auszug aus dem Elternhaus, Heirat, Scheidung, Geburt von Kindern, Todesfälle in der Familie, Pensionierung, usw.)							
Anzahl der Personen in Ihrem Haushalt							
Angaben zu Ihren Wohnorten (weitere Angaben auf Seite 8)							
Nummerierung der Wohnorte							
Angaben zu Ihrem Besitz von Autos und ÖV-Abonnementen							
Verfügbarkeit eines Autos: immer verfügbar							
Verfügbarkeit eines Autos: teilweise verfügbar							
Verfügbarkeit eines Autos: nie verfügbar							
Besitz eines Halbtaxabonnements							
Besitz eines Generalabonnements							
Besitz eines Jahres- oder Monatsabonnements							
Angaben zu Ihrem Ausbildungs- bzw. Arbeitsplatz							
Postleitzahl und Gemeinde des Ausbildungsplatzes							
Postleitzahl und Gemeinde des Arbeitsplatzes							
Meistbenutztes Verkehrsmittel auf dem Weg zum Ausbildungs- bzw. Arbeitsplatz							
Auto, Motorrad, Moped, Mofa							
Eisenbahn, Tram, Bus							
Velo							
zu Fuss							
Persönliches monatliches Brutto-Einkommen (Fremdwährung gegebenenfalls umrechnen)							
unter 2 000 Fr.							
2 000 bis 5 999 Fr.							
6 000 bis 9 999 Fr.							
10 000 bis 13 999 Fr.							
14 000 Fr. und mehr							

Eidgenössische Technische Hochschule Zürich
Swiss Federal Institute of Technology Zurich

1993	1994	1995	1996	1997	1998	1999	2000	2001	2002	2003	2004

Lebensverlaufskalender (Reserveexemplar)
Beachten Sie bitte die Hinweise auf Seite 2.

	1985	1986	1987	1988	1989	1990	1991
Angaben zu Ihrer Familiengeschichte							
Eintragen der Zeitpunkte von wichtigen familiären Ereignissen (z. B. Geburt von Geschwistern, Auszug aus dem Elternhaus, Heirat, Scheidung, Geburt von Kindern, Todesfälle in der Familie, Pensionierung, usw.)							
Anzahl der Personen in Ihrem Haushalt							
Angaben zu Ihren Wohnorten (weitere Angaben auf Seite 8)							
Nummerierung der Wohnorte							
Angaben zu Ihrem Besitz von Autos und ÖV-Abonnementen							
Verfügbarkeit eines Autos: immer verfügbar							
Verfügbarkeit eines Autos: teilweise verfügbar							
Verfügbarkeit eines Autos: nie verfügbar							
Besitz eines Halbtaxabonnements							
Besitz eines Generalabonnements							
Besitz eines Jahres- oder Monatsabonnements							
Angaben zu Ihrem Ausbildungs- bzw. Arbeitsplatz							
Postleitzahl und Gemeinde des Ausbildungsplatzes							
Postleitzahl und Gemeinde des Arbeitsplatzes							
Meistbenutztes Verkehrsmittel auf dem Weg zum Ausbildungs- bzw. Arbeitsplatz							
Auto, Motorrad, Moped, Mofa							
Eisenbahn, Tram, Bus							
Velo							
zu Fuss							
Persönliches monatliches Brutto-Einkommen (Fremdwährung gegebenenfalls umrechnen)							
unter 2 000 Fr.							
2 000 bis 5 999 Fr.							
6 000 bis 9 999 Fr.							
10 000 bis 13 999 Fr.							
14 000 Fr. und mehr							

1993	1994	1995	1996	1997	1998	1999	2000	2001	2002	2003	2004

ETH
Eidgenössische Technische Hochschule Zürich
Swiss Federal Institute of Technology Zurich

Teil 2: Personenfragebogen

Bitte machen Sie hier weitere Angaben zu Ihren (wichtigsten) Wohnorten seit 1985.

1. Wohnort

Frage	Antwort
Bitte geben Sie die Adresse (Postleitzahl, Gemeinde, Strasse und Hausnummer) des Wohnortes an.	
Weshalb sind Sie dorthin gezogen? (Mehrfachnennungen möglich)	☐ familiäre Gründe ☐ berufliche Gründe ☐ Wohnungsgründe ☐ Wohnumfeldgründe ☐ Nähe zur Familie und zu Freunden ☐ Sonstiges ____
Bitte geben Sie die Art Ihres Haushaltes nach dem Umzug an.	☐ Ein-Personen-Haushalt ☐ Familienhaushalt/Paare ohne Kinder ☐ Nicht-Familienhaushalt
Haben Sie zur Miete oder in Eigentum gewohnt? Und wie hoch waren Ihre Wohnkosten beim Einzug? Bitte geben Sie die Miete pro Monat oder den Eigenmietwert pro Jahr an.	☐ Miete ____ Fr. pro Monat ☐ Eigenmietwert ____ Fr. pro Jahr
Wie viele bewohnbare Zimmer hatte Ihre Wohnung/Ihr Haus?	____ bewohnbare Zimmer

2. Wohnort

(gleiche Felder wie 1. Wohnort)

3. Wohnort

☐ familiäre Gründe
☐ berufliche Gründe
☐ Wohnungsgründe
☐ Wohnumfeldgründe
☐ Nähe zur Familie und zu Freunden
☐ Sonstiges ____

☐ Ein-Personen-Haushalt
☐ Familienhaushalt/Paare ohne Kinder
☐ Nicht-Familienhaushalt

☐ Miete ____ Fr. pro Monat
☐ Eigenmietwert ____ Fr. pro Jahr

____ bewohnbare Zimmer

4. Wohnort

☐ familiäre Gründe
☐ berufliche Gründe
☐ Wohnungsgründe
☐ Wohnumfeldgründe
☐ Nähe zur Familie und zu Freunden
☐ Sonstiges ____

☐ Ein-Personen-Haushalt
☐ Familienhaushalt/Paare ohne Kinder
☐ Nicht-Familienhaushalt

☐ Miete ____ Fr. pro Monat
☐ Eigenmietwert ____ Fr. pro Jahr

____ bewohnbare Zimmer

5. Wohnort

☐ familiäre Gründe
☐ berufliche Gründe
☐ Wohnungsgründe
☐ Wohnumfeldgründe
☐ Nähe zur Familie und zu Freunden
☐ Sonstiges ____

☐ Ein-Personen-Haushalt
☐ Familienhaushalt/Paare ohne Kinder
☐ Nicht-Familienhaushalt

☐ Miete ____ Fr. pro Monat
☐ Eigenmietwert ____ Fr. pro Jahr

____ bewohnbare Zimmer

Herzlichen Dank für Ihre Unterstützung!

Part 1: Household form

Please fill in the address of your place of residence.

Street and house number	
Post code	Municipality

For each person of your household, please fill in the year of birth, the sex and the current place of education or employment and indicate whether it is a place of education or employment.

Year of birth	Sex		Current place of employment or education		
	Male	Female	Post code	Municipality	
	☐	☐			☐ Education / ☐ Employment
	☐	☐			☐ Education / ☐ Employment
	☐	☐			☐ Education / ☐ Employment
	☐	☐			☐ Education / ☐ Employment
	☐	☐			☐ Education / ☐ Employment
	☐	☐			☐ Education / ☐ Employment
	☐	☐			☐ Education / ☐ Employment
	☐	☐			☐ Education / ☐ Employment
	☐	☐			☐ Education / ☐ Employment

What is the gross income per month of your household?

- ☐ Under 2 000 CHF
- ☐ 2 000 to 3 999 CHF
- ☐ 4 000 to 5 999 CHF
- ☐ 6 000 to 7 999 CHF
- ☐ 8 000 to 9 999 CHF
- ☐ 10 000 to 11 999 CHF
- ☐ 12 000 to 13 999 CHF
- ☐ 14 000 CHF and more

How many vehicles has your household at its disposal?

	Cars	☐ No cars		Motorcycles with more than 125 ccm	☐ No motorcycles
	Operable bicycles	☐ No bicycles		Small motorcycles with less than 125 ccm	☐ No small motorcycles

Part 2: Person form

Eidgenössische Technische Hochschule Zürich
Swiss Federal Institute of Technology Zurich

When were you born?

Month [ ]   Year [ ]

Are you ...

☐ Male   ☐ Female

Are you ...

☐ Swiss national
☐ Other nationality, namely [ ]

Which qualifications have you aquired? (Tick all, which apply)

☐ Primary school   ☐ Apprenticeship   ☐ University of applied sciences degree
☐ Secondary school   ☐ Baccalauréat   ☐ University degree/ETH degree

Are you ... (Tick all, which apply)

☐ Full-time employed   ☐ In education or apprenticeship   ☐ Home duties
☐ Part-time employed   ☐ Job-seeking   ☐ Retired

What type of education or employment are you mainly engaged in?

Education [ ]   Employment [ ]

How many hours are you in education or employment engaged?

Education [ ] Hours   Employment [ ] Hours

What is the exact address of your current place of education or employment?
Please indicate post code, municipality, street and house number.

Education [ ]   Employment [ ]

How long do you need or would you need for the trip (from door to door)
to your place of education or employment if you were to use ...

	Trip to education		Trip to employment	
Only a car?	Circa [ ]	Minutes	Circa [ ]	Minutes
Only public transport?	Circa [ ]	Minutes	Circa [ ]	Minutes
Trip is not possible with public transport.	☐		☐	

Do you own a driving licence for cars? If yes, since when?

☐ No   ☐ Yes   Year of acquisition [ ]

Part 2: Person form

How often is a car available to you?

☐ Always ☐ Frequently ☐ Infrequently ☐ Never ☐ Via car sharing

How satisfied are you ...

	Very satisfied									Very dissatisfied
	1	2	3	4	5	6	7	8	9	10
With your health?	☐	☐	☐	☐	☐	☐	☐	☐	☐	☐
With your accommodation?	☐	☐	☐	☐	☐	☐	☐	☐	☐	☐
With your work?	☐	☐	☐	☐	☐	☐	☐	☐	☐	☐
With your leisure time?	☐	☐	☐	☐	☐	☐	☐	☐	☐	☐
With the condition of the environment in your region?	☐	☐	☐	☐	☐	☐	☐	☐	☐	☐
With your life overall?	☐	☐	☐	☐	☐	☐	☐	☐	☐	☐

Will you move within the next year?

☐ Very likely ☐ Likely ☐ Unlikely ☐ Very unlikely

Questions about your residential history

Please fill in the following life course calendar for the time between 1985 and 2004.
Please note:

- There is an **example** of a filled in life course calendar on the next page.
- On the following pages you find two life course calendars. Please fill in **one** of them. The second one is meant as a reserve exemplar in case, that corrections make the first one too difficult to read and understand. In this case please cross out the first one clearly.
- It might be easier, if you start by entering important events of the family history (e.g. birth of siblings, moving out of your parents' house, marriage, divorce, birth of children, deaths in the family, retirement, etc.).
  It might also help to start with later events and then proceed backwards.
- The main interest of this survey is your residential history during the last twenty years. Therefore please mark your moves clearly, number the places of residence and answer the further questions for the different places of residence on page 8. In case the five given places there are not sufficient, please choose the five most important places of residence and note down the number of the place of residence from the life course calendar.
- The questions concerning the different places of residence also include the exact address, preferably as post code, municipality, street and house number. If you can not remember the address exactly, please enter the name of the municipality and in case of bigger towns the name of the district.
- If you were living abroad during the time between 1985 and 2004, we are also interested in the information about your residences there.

When did you move the last time before 1985?

| | Year of the last move before 1985 |

☐ I did not move before 1985.

When did you change your place of education or employment the last time before 1985?

| | Year of the last change of the place of education or employment before 1985 |

☐ I did not change the place of education or employment before 1985.
☐ I was not engaged in any education or employment before 1985.

Part 2: Person form

Example of a filled in life course calendar

	1985	1986	1987	1988	1989	1990	1991	1992
Information about your family history								
Please indicate important family events (e.g. birth of siblings, moving out of your parents' house, marriage, divorce, birth of children, deaths in the family, retirement, etc.)			⊢moving out of the parents' house			⊢marriage⊢birth of the 1st child		⊢birth of the
Number of persons in your household	⊢— 4 —⊢		— 1 —⊢		— 2 —⊢		3 —⊢	— 4 —
Information about your places of residence (Please fill in further information on page 8)								
Please number the places of residence	⊢—1st place—⊢		—2nd place—⊢		—3rd place—⊢			⊢—4th place
Information about your ownership of cars and public transport tickets								
Availability of a car: always available						⊢		
Availability of a car: partially available	⊢	—⊢						
Availability of a car: never available			⊢	—⊢				
Ownership of a half-fare discount ticket	⊢							
Ownership of a national annual ticket								
Ownership of a regional annual or monthly ticket	⊢	—⊢						
Information about your places of education and employment								
Post code and municipality of the place of education	⊢8090 Zurich⊢							
Post code and municipality of the place of employment			⊢— 8001 Zurich —⊢				⊢—8303 Bassersdorf —	
Mostly used mode of transport for the trip to your place of education or employment								
Car, motorcycle, moped						⊢		
Train, tram, bus		⊢	—⊢					
Bicycle			⊢		—⊢			
On foot								
Personal gross income per month (Please convert foreign currency if necessary)								
Under 2 000 CHF	⊢	—⊢						
2 000 to 5 999 CHF			⊢		—⊢			
6 000 to 9 999 CHF						⊢		
10 000 to 13 999 CHF								
14 000 CHF and more								

Life course calendar
Please consider the hints on page 2.

	1985	1986	1987	1988	1989	1990	1991
Information about your family history							
Please indicate important family events (e.g. birth of siblings, moving out of your parents' house, marriage, divorce, birth of children, deaths in the family, retirement, etc.)							
Number of persons in your household							
Information about your places of residence (Please fill in further information on page 8)							
Please number the places of residence							
Information about your ownership of cars and public transport tickets							
Availability of a car: always available							
Availability of a car: partially available							
Availability of a car: never available							
Ownership of a half-fare discount ticket							
Ownership of a national annual ticket							
Ownership of a regional annual or monthly ticket							
Information about your places of education and employment							
Post code and municipality of the place of education							
Post code and municipality of the place of employment							
Mostly used mode of transport for the trip to your place of education or employment							
Car, motorcycle, moped							
Train, tram, bus							
Bicycle							
On foot							
Personal gross income per month (Please convert foreign currency if necessary)							
Under 2 000 CHF							
2 000 to 5 999 CHF							
6 000 to 9 999 CHF							
10 000 to 13 999 CHF							
14 000 CHF and more							

1993	1994	1995	1996	1997	1998	1999	2000	2001	2002	2003	2004

Life course calendar   (reserve exemplar)
Please consider the hints on page 2.

	1985	1986	1987	1988	1989	1990	1991
Information about your family history							
Please indicate important family events (e.g. birth of siblings, moving out of your parents` house, marriage, divorce, birth of children, deaths in the family, retirement, etc.)							
Number of persons in your household							
Information about your places of residence (Please fill in further information on page 8)							
Please number the places of residence							
Information about your ownership of cars and public transport tickets							
Availability of a car: always available							
Availability of a car: partially available							
Availability of a car: never available							
Ownership of a half-fare discount ticket							
Ownership of a national annual ticket							
Ownership of a regional annual or monthly ticket							
Information about your places of education and employment							
Post code and municipality of the place of education							
Post code and municipality of the place of employment							
Mostly used mode of transport for the trip to your place of education or employment							
Car, motorcycle, moped							
Train, tram, bus							
Bicycle							
On foot							
Personal gross income per month (Please convert foreign currency if necessary)							
Under 2 000 CHF							
2 000 to 5 999 CHF							
6 000 to 9 999 CHF							
10 000 to 13 999 CHF							
14 000 CHF and more							

1993	1994	1995	1996	1997	1998	1999	2000	2001	2002	2003	2004

Part 2: Person form

Please fill in further information about your (most important) places of residence since 1985.

1st place of residence

Please fill in the address (post code, municipality, street and house number) of the place of residence.

Why did you move there? (Tick all, which apply)
- [ ] Family reasons
- [ ] Work related reasons
- [ ] Accommodation related reasons
- [ ] Quality of surrounding environment
- [ ] Vicinity to family and friends
- [ ] Other

What type of household did you live in after the move.
- [ ] Single-person household
- [ ] Family/Couples without children
- [ ] Non-family household

Did you rent or own the accommodation? And how high were the costs? Please enter the rent per month or the rental value for property per year.
- [ ] Rent _____ CHF per month
- [ ] Rental value _____ CHF per year

How many inhabitable rooms had the place of residence? _____ Inhabitable rooms

2nd place of residence

- [ ] Family reasons
- [ ] Work related reasons
- [ ] Accommodation related reasons
- [ ] Quality of surrounding environment
- [ ] Vicinity to family and friends
- [ ] Other

- [ ] Single-person household
- [ ] Family/Couples without children
- [ ] Non-family household

- [ ] Rent _____ CHF per month
- [ ] Rental value _____ CHF per year

_____ Inhabitable rooms

3rd place of residence

- [ ] Family reasons
- [ ] Work related reasons
- [ ] Accommodation related reasons
- [ ] Quality of surrounding environment
- [ ] Vicinity to family and friends
- [ ] Other

- [ ] Single-person household
- [ ] Family/Couples without children
- [ ] Non-family household

- [ ] Rent _____ CHF per month
- [ ] Rental value _____ CHF per year

_____ Inhabitable rooms

4th place of residence

- [ ] Family reasons
- [ ] Work related reasons
- [ ] Accommodation related reasons
- [ ] Quality of surrounding environment
- [ ] Vicinity to family and friends
- [ ] Other

- [ ] Single-person household
- [ ] Family/Couples without children
- [ ] Non-family household

- [ ] Rent _____ CHF per month
- [ ] Rental value _____ CHF per year

_____ Inhabitable rooms

5th place of residence

- [ ] Family reasons
- [ ] Work related reasons
- [ ] Accommodation related reasons
- [ ] Quality of surrounding environment
- [ ] Vicinity to family and friends
- [ ] Other

- [ ] Single-person household
- [ ] Family/Couples without children
- [ ] Non-family household

- [ ] Rent _____ CHF per month
- [ ] Rental value _____ CHF per year

_____ Inhabitable rooms

Thank you very much for your assistance!

Appendix B: Models for the mobility tool ownership

B 1 Models for the mobility tool ownership for the year 2005

Figure B 1-1 Utility for age and gender in the binomial logit models for car availability and public transport season ticket ownership (2005)

Car availability: Always

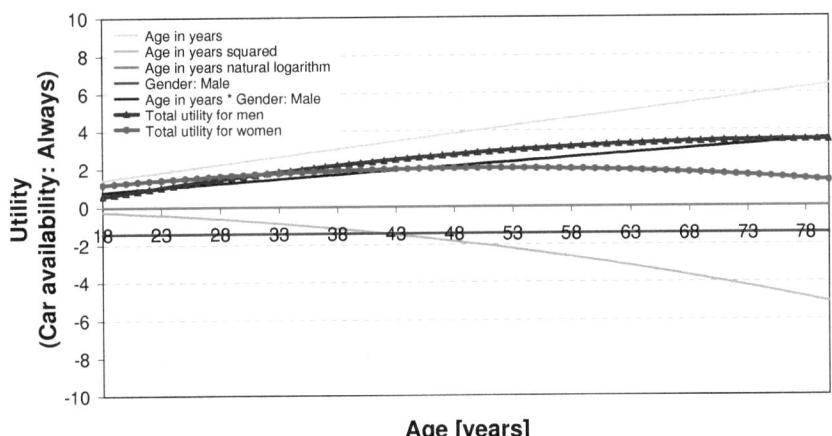

Car availability: Partially

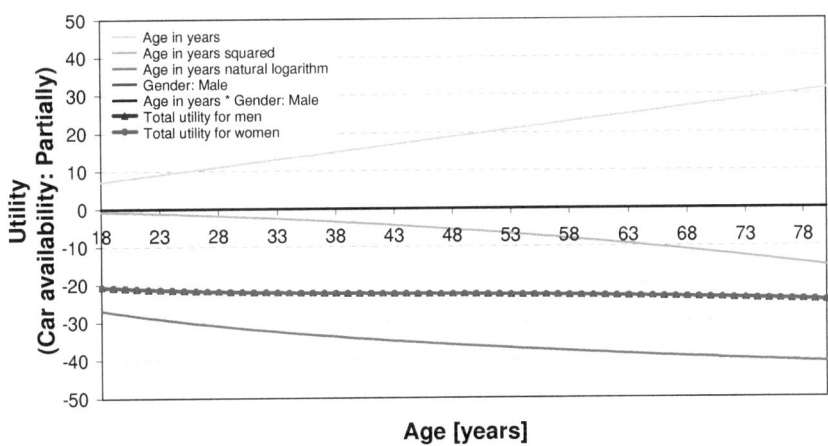

Figure B 1-1 is continued ...

Figure B 1-1 continued ...

National ticket ownership

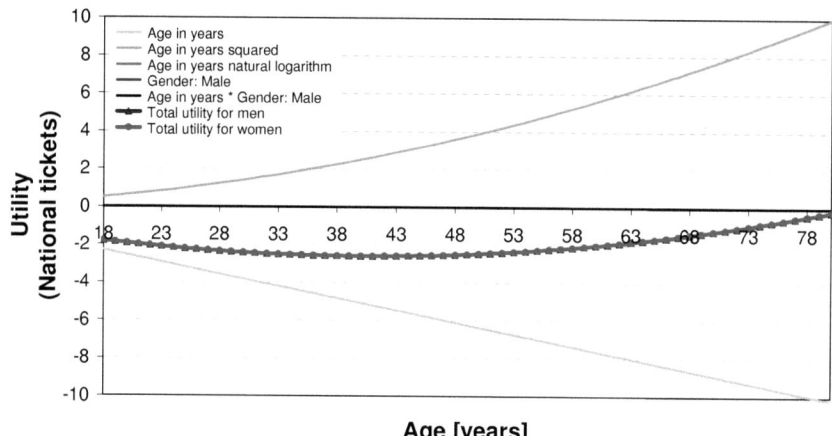

Regional ticket ownership

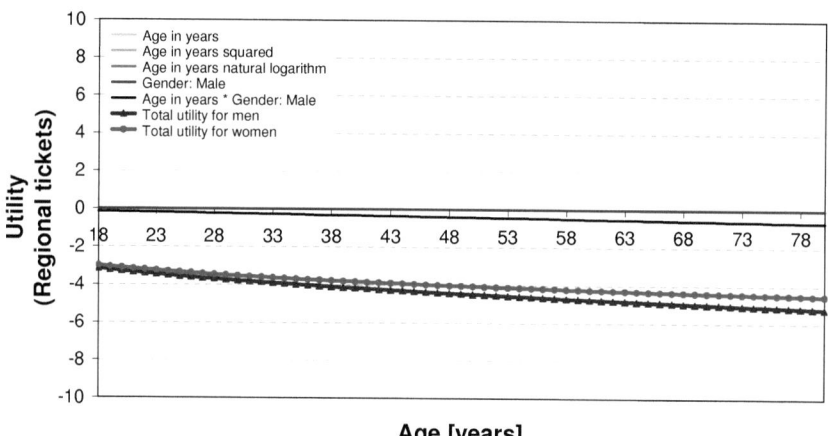

Figure B 1-1 is continued ...

Figure B 1-1 continued ...

Half-fare discount ticket ownership

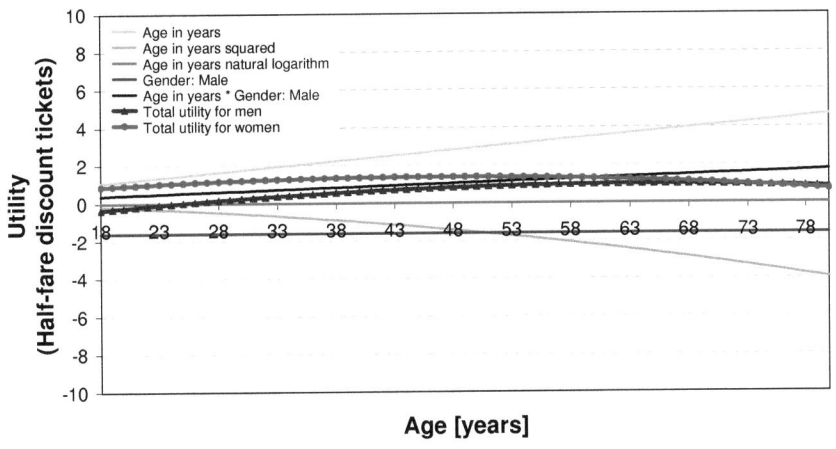

Figure B 1-2 Utility for the monthly income in the binomial logit models for car availability and public transport season ticket ownership (2005)

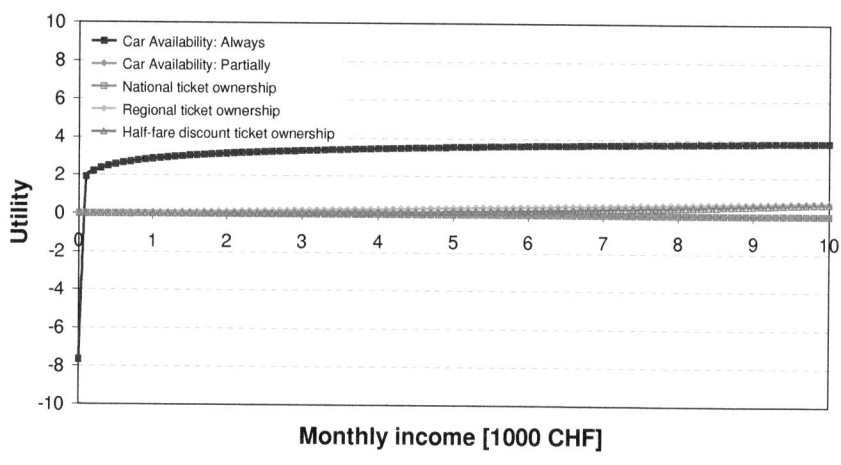

Figure B 1-3 Utility for the accommodation costs in the binomial logit models for car availability and public transport season ticket ownership (2005)

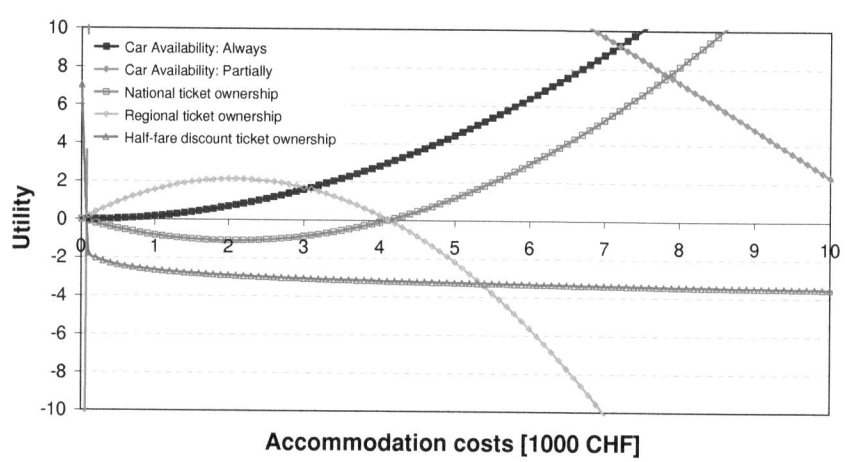

Figure B 1-4   Utility for age and gender in the binomial logit models for car availability and public transport season ticket ownership only considering persons in education and employment (2005)

Car availability: Always

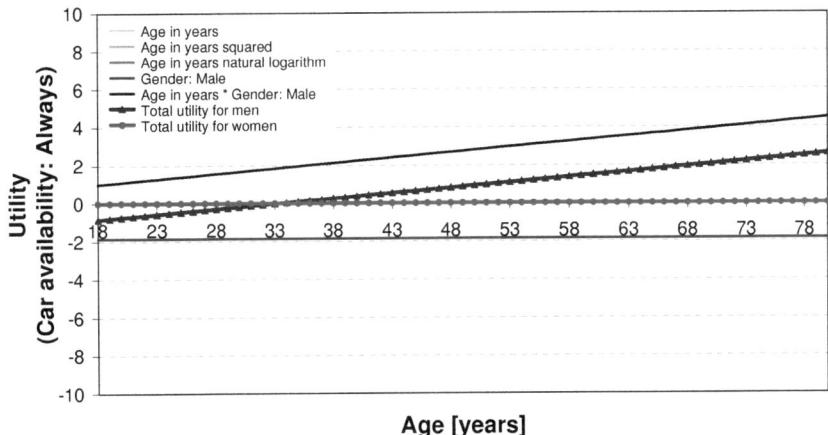

Car availability: Partially

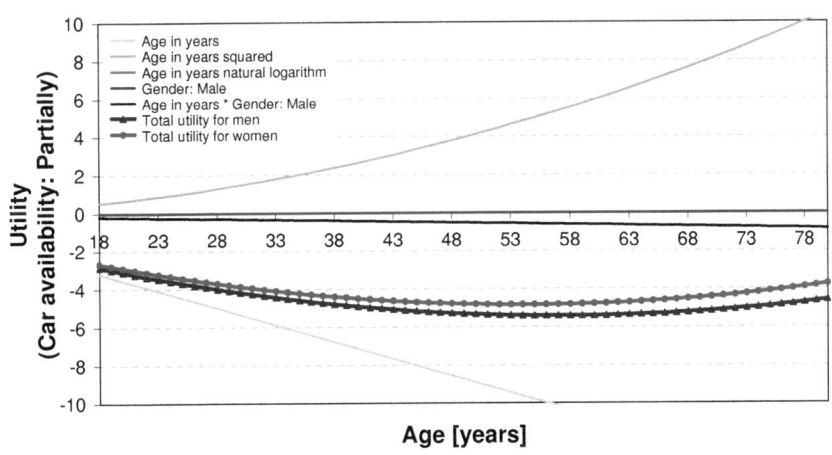

Figure B 1-4 is continued ...

Figure B 1-4 continued ...

National ticket ownership

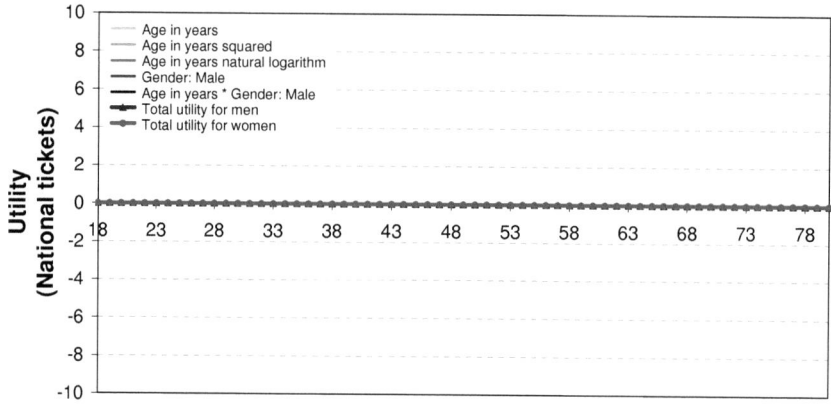

Regional ticket ownership

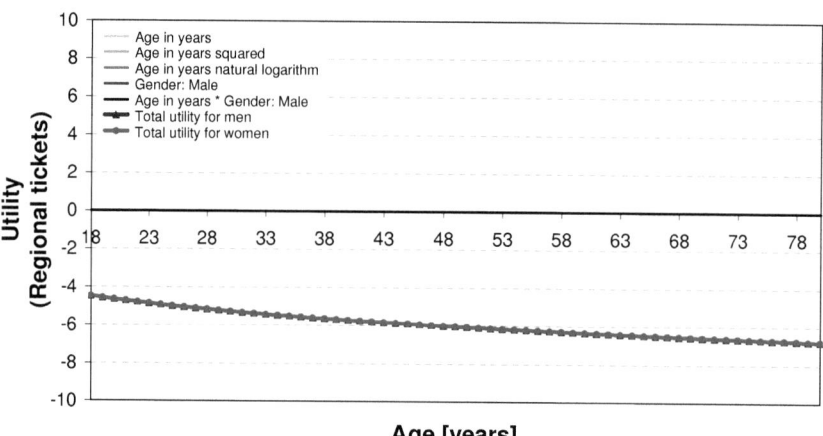

Figure B 1-4 is continued ...

Figure B 1-4 continued ...

Half-fare discount ticket ownership

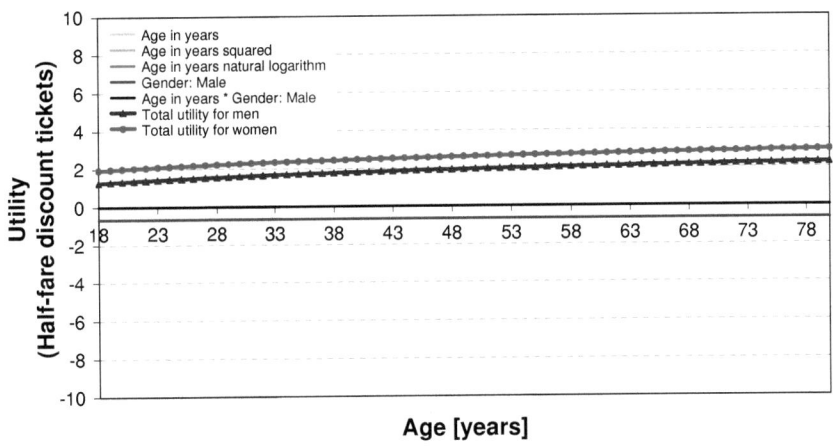

Figure B 1-5 Utility for the monthly income in the binomial logit models for car availability and public transport season ticket ownership only considering persons in education and employment (2005)

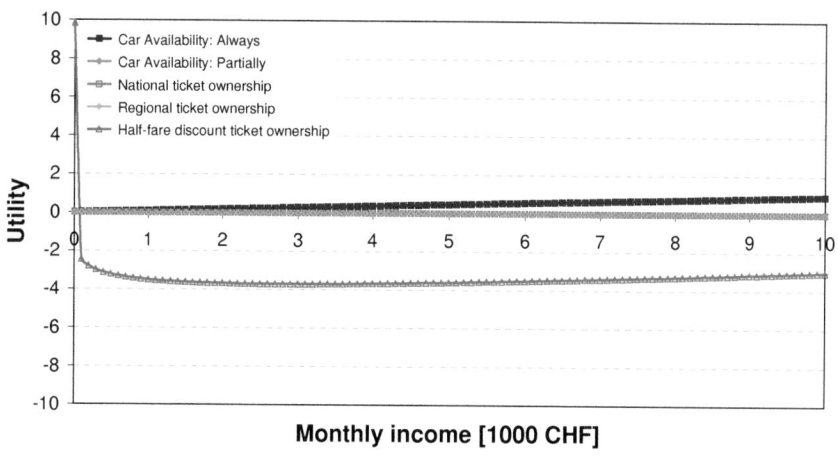

Figure B 1-6 Utility for the accommodation costs in the binomial logit models for car availability and public transport season ticket ownership only considering persons in education and employment (2005)

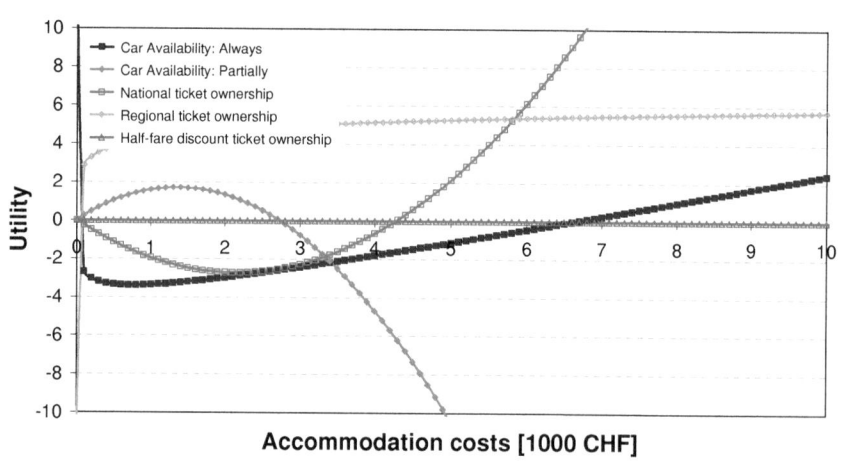

Table B 1-1 Nested logit model for mobility tool ownership in groups with two nests for car and no car (2005)

Explanatory variable	No Car + No Tickets	No Car + HF T	No Car + Nat T / Reg T	Car + No Tickets	Car + HF T	Car + Nat T / Reg T
Age in years	– 0.961		– 0.745			– 1.126
Age in years squared	+ 0.007		+ 0.005			+ 0.008
Age in years natural logarithm	+ 13.788	+ 2.878	+ 13.053		+ 4.120	+ 14.483
Gender: Male		– 2.120			– 2.710	– 2.077
Age in years * Gender: Male	– 0.052		– 0.052			
Nationality: Swiss national		+ 2.287	+ 2.030		+ 5.027	+ 1.971
College or university degree		+ 2.341	+ 1.964		+ 2.876	+ 3.004
In education	+ 3.366	+ 3.487	+ 5.537		+ 4.255	+ 8.077
In employment						+ 1.468
Monthly income in 1000 CHF			+ 1.072		+ 0.673	+ 0.743
Monthly income in 1000 CHF squared			– 0.020			
Monthly income natural logarithm	– 1.592		– 2.365		– 1.371	– 2.263
Accommodation costs natural logarithm				+ 1.456		
Spatial and transport system defined classification:						
Main centres (referential category)						
Middle and ancillary centres with railway access				+ 3.131		
Middle and ancillary centres without railway access				+ 2.536		
Agglomeration municipalities				+ 2.129		
Rural areas				+ 4.331		

Model parameters for the two nests:	
Nest: Car	0.248 *
Nest: No Car	0.662 *
Number of observations	1154
ρ^2 (adjusted)	0.211

For all variables: Level of significance ≤ 0.10

* For parameters: Level of significance ≤ 0.10

Table B 1-2   Nested logit model for mobility tool ownership in groups with two nests for national and regional tickets and no national and regional tickets (2005)

Explanatory variable	No Car + No Tickets	No Car + HF T	No Car + Nat T / Reg T	Car + No Tickets	Car + HF T	Car + Nat T / Reg T
Age in years	− 1.451		− 0.600		+ 0.050	− 0.114
Age in years squared	+ 0.012	+ 0.001	+ 0.006			
Age in years natural logarithm	+ 18.496		+ 10.595			
Gender: Male	+ 3.540		+ 3.589		− 1.606	− 3.883
Age in years * Gender: Male	− 0.116	− 0.048	− 0.142			+ 0.092
Nationality: Swiss national	− 2.417	+ 1.635			+ 2.859	+ 2.139
College or university degree		+ 1.634			+ 1.814	+ 2.403
In education					+ 2.964	+ 4.553
In employment						+ 0.810
Monthly income in 1000 CHF	+ 7.348		+ 1.727		+ 0.366	
Monthly income in 1000 CHF squared	− 1.223		− 0.043			
Monthly income natural logarithm	− 5.014	− 0.558	− 4.186		− 0.533	
Accommodation costs in 1000 CHF per month				+ 1.124		
Accommodation costs in 1000 CHF squared				− 0.190		
Spatial and transport system defined classification:						
Main centres (referential category)						
Middle and ancillary centres with railway access				+ 1.967		
Middle and ancillary centres without railway access				+ 1.919		
Agglomeration municipalities				+ 1.666		
Rural areas				+ 2.978		

Model parameters for the two nests:	
Nest: National and regional tickets	0.121 *
Nest: No National and regional tickets	0.421 *
Number of observations	1154
ρ^2 (adjusted)	0.210

For all variables: Level of significance ≤ 0.10
* For parameters: Level of significance ≤ 0.10

Table B 1-3 Nested logit model for mobility tool ownership in groups with two nests for half-fare discount tickets and no half-fare discount tickets (2005)

Explanatory variable	No Car + No Tickets	No Car + HF T	No Car + Nat T / Reg T	Car + No Tickets	Car + HF T	Car + Nat T / Reg T
Age in years	− 0.108	+ 1.056	− 0.067		− 0.032	− 0.087
Age in years squared	+ 0.001	− 0.006	+ 0.001			+ 0.001
Age in years natural logarithm	+ 0.870	− 17.374	+ 0.782		+ 1.463	+ 0.961
Gender: Male		− 3.917			− 1.222	− 0.173
Age in years * Gender: Male	− 0.006		− 0.006		+ 0.034	
Nationality: Swiss national	− 0.340				+ 1.230	
College or university degree			+ 0.179		+ 0.669	+ 0.246
In education			+ 0.450			+ 0.548
In employment		− 4.702			+ 1.364	+ 0.195
Monthly income in 1000 CHF	− 0.131		+ 0.056		+ 0.396	+ 0.059
Monthly income in 1000 CHF squared		− 0.087				
Monthly income natural logarithm		+ 2.879	− 0.161		− 1.297	− 0.179
Accommodation costs in 1000 CHF per month				+ 0.103		
Spatial and transport system defined classification:						
Main centres (referential category)						
Middle and ancillary centres with railway access				+ 0.287		
Middle and ancillary centres without railway access				+ 0.268		
Agglomeration municipalities				+ 0.245		
Rural areas				+ 0.453		

Model parameters for the two nests:	
Nest: Half-fare discount tickets	0.107 *
Nest: No Half-fare discount tickets	4.402
Number of observations	1154
ρ^2 (adjusted)	0.206

For all variables: Level of significance ≤ 0.10

* For parameters: Level of significance ≤ 0.10

Table B 1-4 Multivariate probit model for mobility tool ownership in groups using all variables for all alternatives (2005)

Explanatory variable	No Car + No Tickets	No Car + HF T	No Car + Nat T / Reg T	Car + No Tickets	Car + HF T	Car + Nat T / Reg T
Age in years	− 0.198 *	+ 0.127	− 0.088 *		+ 0.089 *	− 0.090
Age in years squared	+ 0.002 *	− 0.001	+ 0.001 *		− 0.001 *	+ 0.001 *
Age in years natural logarithm	+ 2.095	− 1.906	+ 1.078		− 0.807	+ 0.173
Gender: Male	+ 0.723	− 0.280	+ 0.553		− 0.651 *	− 0.412
Age in years * Gender: Male	− 0.017	+ 0.003	− 0.016 *		+ 0.012 *	+ 0.011
Nationality: Swiss national	− 0.893 *	+ 0.090	− 0.008		+ 0.613 *	+ 0.093
College or university degree	− 0.228	+ 0.194	+ 0.038		+ 0.231 *	+ 0.303 *
In education	− 0.485	− 0.401	+ 0.338		− 0.130	+ 0.794 *
In employment	+ 0.009	− 0.154	− 0.105		− 0.027	+ 0.406 *
Monthly income in 1000 CHF	+ 0.695	− 0.058	+ 0.284 *		+ 0.055	+ 0.012
Monthly income in 1000 CHF squared	− 0.132	− 0.005	− 0.012 *		+ 0.000	− 0.001
Monthly income natural logarithm	− 0.545	+ 0.254	− 0.468 *		− 0.103	+ 0.020
Accommodation costs in 1000 CHF				+ 0.856 *		
Accommodation costs in 1000 CHF squared				− 0.186 *		
Accommodation costs natural logarithm				− 0.219 *		
Spatial and transport system defined classification:						
Main centres (referential category)						
Middle and ancillary centres with railway access				+ 0.382		
Middle and ancillary centres without railway access				+ 0.570 *		
Agglomeration municipalities				+ 0.462		
Rural areas				+ 0.793 *		
Population in the residential municipality in 1000 inhabitants				+ 0.001		
Population density in the residential municipality in inhabitants per square kilometre				− 0.012		
Correlation matrix:						
No Car + No Tickets	+ 1.000	− 0.054	− 0.085	− 0.156	− 0.129	− 0.106
No Car + HF T		+ 1.000	− 0.090	− 0.172	− 0.138	− 0.117
No Car + Nat T / Reg T			+ 1.000	− 0.268	− 0.219	− 0.184
Car + No Tickets				+ 1.000	− 0.411 *	− 0.336 *
Car + HF T					+ 1.000	− 0.281 *
Car + Nat T / Reg T						+ 1.000
Number of observations						1154
ρ^2 (adjusted)						0.503

* Level of significance ≤ 0.10

Table B 1-5 Multivariate probit model for mobility tool ownership in groups using the variables of the corresponding multinomial logit model (2005)

Explanatory variable	No Car + No Tickets	No Car + HF T	No Car + Nat T / Reg T	Car + No Tickets	Car + HF T	Car + Nat T / Reg T
Age in years	− 0.200 *		− 0.093 *		+ 0.088 *	− 0.078 *
Age in years squared	+ 0.002 *	− 0.000 *	+ 0.001 *		− 0.001 *	+ 0.001 *
Age in years natural logarithm	+ 2.327		+ 1.095		− 0.716 *	+ 0.120
Gender: Male	+ 0.670		+ 0.547 *			
Age in years * Gender: Male	− 0.017 *	− 0.012 *	− 0.018 *		+ 0.002 *	+ 0.002
Nationality: Swiss national	− 0.930 *				+ 0.605 *	+ 0.083
College or university degree		− 0.721 *	+ 0.027		+ 0.222 *	+ 0.302 *
In education			+ 0.335 *		− 0.172 *	+ 0.763 *
In employment						+ 0.402 *
Monthly income in 1000 CHF	+ 0.949		+ 0.281 *		+ 0.059 *	− 0.010
Monthly income in 1000 CHF squared	− 0.147 *		− 0.012 *			
Monthly income natural logarithm	− 0.738		− 0.482 *		− 0.170 *	+ 0.008
Accommodation costs in 1000 CHF				+ 0.873 *		
Accommodation costs in 1000 CHF squared				− 0.195 *		
Accommodation costs natural logarithm				− 0.232 *		
Spatial and transport system defined classification:						
Main centres (referential category)						
Middle and ancillary centres with railway access				+ 0.398 *		
Middle and ancillary centres without railway access				+ 0.525 *		
Agglomeration municipalities				+ 0.477 *		
Rural areas				+ 0.857 *		
Correlation matrix:						
No Car + No Tickets	+ 1.000	− 0.040 *	− 0.066	− 0.137	− 0.106	− 0.081
No Car + HF T		+ 1.000	− 0.085	− 0.178	− 0.154	− 0.105
No Car + Nat T / Reg T			+ 1.000	− 0.246	− 0.226	− 0.171
Car + No Tickets				+ 1.000	− 0.458 *	− 0.333
Car + HF T					+ 1.000	− 0.289
Car + Nat T / Reg T						+ 1.000
Number of observations						1154
ρ^2 (adjusted)						0.477

* Level of significance ≤ 0.10

B 2 Models for the mobility tool ownership for the period from 1985 to 2004

Figure B 2-1 Personal and familial events by gender, age and birth cohort membership (1985-2004)

Moving out of parents' house

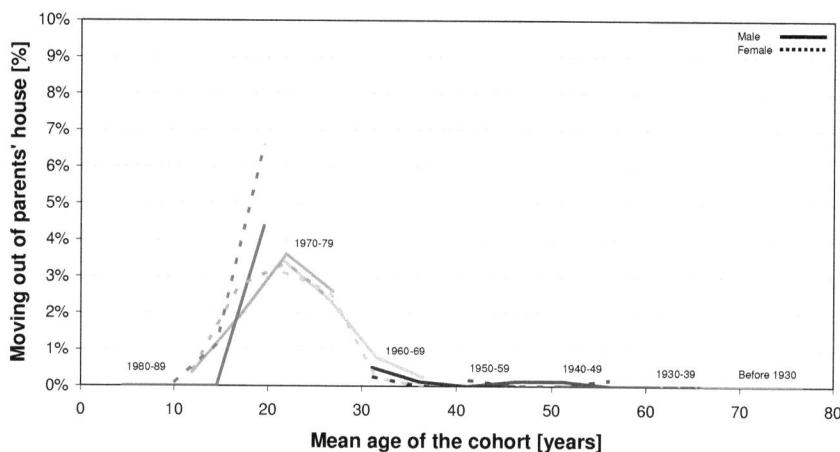

Birth of persons in the household

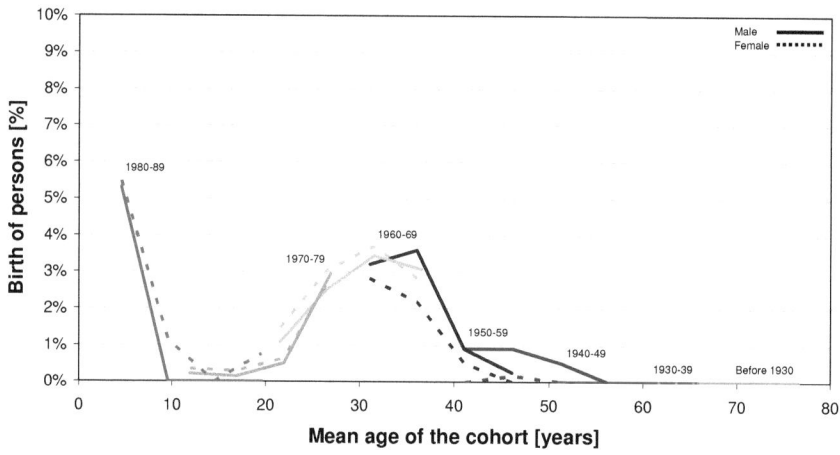

Figure B 2-1 is continued ...

Figure B 2-1 continued ...

Partnership / marriage

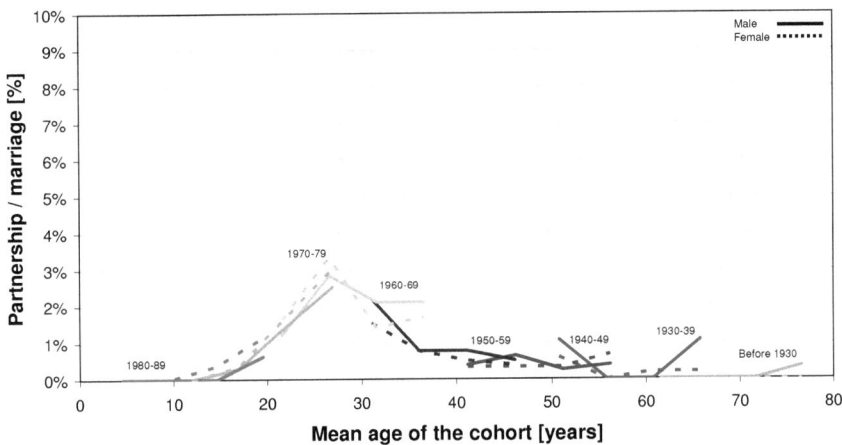

Break-up / divorce

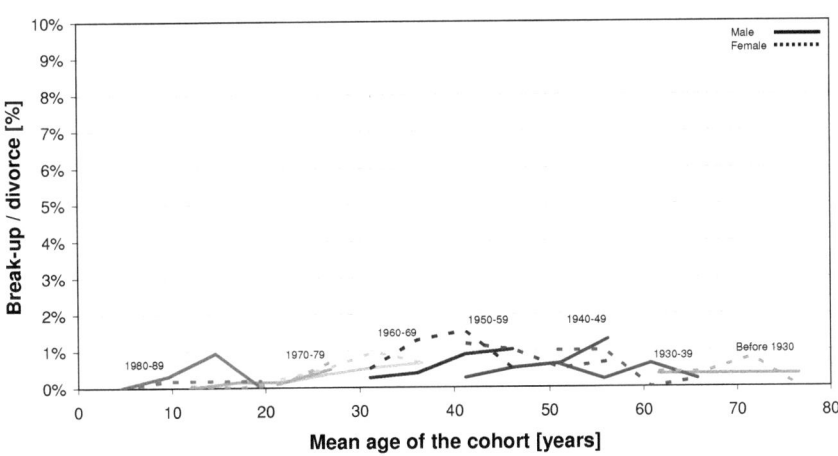

Table B 2-1   Binomial logit models for car availability and public transport season ticket ownership not taking the panel effect into account (1985-2004)

Explanatory variable	Car availability: Always	Car availability: Partially	National ticket ownership	Regional ticket ownership	Half-fare discount ticket ownership
Age in years	+ 0.076 *	+ 0.009	– 0.102 *	– 0.042 *	+ 0.059 *
Age in years squared	– 0.001 *	– 0.001 *	+ 0.001 *	+ 0.000 *	– 0.000 *
Gender: Male	+ 0.790 *	– 0.431 *	+ 0.284 *	– 0.237 *	– 0.378 *
Nationality: Swiss national	+ 0.213 *	+ 0.681 *	+ 1.556 *	– 0.241 *	+ 1.023 *
College or university degree	+ 0.102 *	+ 0.348 *	+ 0.564 *	+ 0.193 *	+ 0.555 *
In education	– 0.064	– 0.048	+ 0.686 *	+ 0.349 *	+ 0.325 *
Change in education	– 0.441 *	+ 0.202 *	+ 0.046	– 0.013	+ 0.114
Distance between the place of residence and the place of education in 1000 kilometres	+ 0.385 *	– 0.057	+ 2.711 *	– 0.071	– 1.141
In employment	+ 0.668 *	– 0.242 *	+ 0.170 *	+ 0.347 *	+ 0.217 *
Change in employment	– 0.017	+ 0.221 *	+ 0.003	+ 0.156 *	+ 0.077
Distance between the place of residence and the place of employment in 1000 kilometres	– 0.028	+ 0.036	+ 0.090	– 3.733 *	– 0.076
Monthly income in 1000 CHF	+ 0.355 *	– 0.234 *	+ 0.163 *	+ 0.063 *	+ 0.024 *
Monthly income in 1000 CHF squared	– 0.013 *	+ 0.009 *	– 0.009 *	– 0.002	+ 0.002 *
Car availability: Always			– 1.937 *	– 1.104 *	– 0.908 *
Car availability: Partially			– 0.485 *	– 0.129 *	– 0.394 *
National ticket ownership	– 2.934 *	+ 0.839 *			
Regional ticket ownership	– 1.423 *	+ 0.585 *			+ 0.523 *
Half-fare discount ticket ownership	– 1.304 *	+ 0.298 *		+ 0.519 *	
Moving out of parents' house	+ 0.277 *	+ 0.183	– 0.034	– 0.003	+ 0.121
Change in residence	+ 0.077	– 0.048	+ 0.140	– 0.055	+ 0.030
Birth of a person in the household	+ 0.245 *	+ 0.121	– 0.187	– 0.195	+ 0.084
Number of persons in the household	– 0.197 *	+ 0.235 *	– 0.186 *	+ 0.033 *	+ 0.001
Number of rooms in the accommodation	+ 0.132 *	– 0.025 *	+ 0.131 *	– 0.076 *	+ 0.006
Degree of urbanisation: Urban (referential category)					
Urban to rural	+ 0.670 *	+ 0.064	– 0.492 *	– 0.385 *	– 0.084 *
Rural	+ 0.503 *	+ 0.265 *	– 0.093	– 1.117 *	– 0.188 *
Place of residence abroad	+ 0.392 *	+ 0.136	– 1.391 *	– 1.372 *	– 0.552 *
Purchasing power index in the residential region	+ 0.007 *	+ 0.014 *	+ 0.040 *	+ 0.008 *	+ 0.010 *
Constant	– 3.439 *	– 3.171 *	– 5.806 *	– 0.145	– 3.645 *

Table B 2-1 is continued ...

Table B 2-1 continued ...

Explanatory variable	Car availability: Always	Car availability: Partially	National ticket ownership	Regional ticket ownership	Half-fare discount ticket ownership
Number of persons	1043	1043	1043	1043	1043
Number of observations	28808	28808	28808	28808	28808
ρ^2 (adjusted)	0.324	0.492	0.680	0.374	0.112

* Level of significance ≤ 0.10

Table B 2-2 Binomial logit models for car availability and public transport season ticket ownership taking the panel effect into account (1985-2004)

Explanatory variable	Car availability: Always	Car availability: Partially	National ticket ownership	Regional ticket ownership	Half-fare discount ticket ownership
Age in years	+ 0.538 *	– 0.089	– 0.474 *	– 0.140 *	+ 0.128 *
Age in years squared	– 0.005 *	– 0.000	+ 0.006 *	+ 0.001	– 0.001
Gender: Male	+ 3.595 *	– 0.842 *	+ 0.972 *	– 0.114	– 1.710 *
Nationality: Swiss national	– 0.147	+ 1.724 *	+ 4.309 *	– 0.053	+ 3.560 *
College or university degree	– 0.245	+ 2.021 *	+ 2.848 *	+ 1.609 *	+ 2.590 *
In education	+ 0.000	– 0.757 *	+ 0.708 *	+ 1.119 *	+ 0.562 *
Change in education	– 0.340 *	+ 0.096	+ 0.186	+ 0.138	+ 0.185
Distance between the place of residence and the place of education in 1000 kilometres	+ 0.227	– 0.145	+ 5.479	+ 0.468	– 3.810
In employment	+ 1.063 *	– 0.317	– 0.109	+ 1.109 *	+ 0.282
Change in employment	+ 0.069	+ 0.181	+ 0.017	+ 0.167 *	+ 0.120
Distance between the place of residence and the place of employment in 1000 kilometres	– 0.493 *	– 0.217	+ 1.222 *	– 8.174 *	– 1.364 *
Monthly income in 1000 CHF	+ 0.511 *	– 0.345 *	+ 0.030	– 0.111	+ 0.228 *
Monthly income in 1000 CHF squared	– 0.024 *	+ 0.014 *	– 0.000	+ 0.011	– 0.012 *
Car availability: Always			– 2.243 *	– 2.370 *	– 1.720 *
Car availability: Partially			– 0.927 *	– 0.120	– 1.039 *
National ticket ownership	– 3.799 *	+ 0.486			
Regional ticket ownership	– 3.076 *	+ 1.220 *			+ 1.080 *
Half-fare discount ticket ownership	– 2.173 *	+ 0.096		+ 0.878 *	
Moving out of parents' house	+ 0.408	+ 0.062	+ 0.061	+ 0.020	+ 0.118
Change in residence	– 0.015	– 0.054	+ 0.212	– 0.195 *	– 0.007
Birth of a person in the household	+ 0.340	+ 0.119	– 0.567	– 0.073	– 0.269
Number of persons in the household	– 0.286 *	+ 0.150	– 0.163	– 0.131	– 0.009
Number of rooms in the accommodation	+ 0.078	+ 0.038	– 0.050	+ 0.065	+ 0.006
Degree of urbanisation: Urban (referential category)					
Urban to rural	– 0.115	+ 0.479 *	– 0.436	– 0.572 *	+ 0.152
Rural	+ 0.933 *	– 0.379	– 1.863 *	– 0.989 *	+ 0.611
Place of residence abroad	+ 0.573	– 0.395	– 2.705 *	– 2.104 *	– 3.002 *
Purchasing power index in the residential region	– 0.003	+ 0.028 *	+ 0.116 *	+ 0.028 *	+ 0.025 *
Constant	– 11.866 *	– 6.331 *	– 15.689 *	– 3.048 *	– 10.609 *
Individual-specific random parameter	+ 6.999 *	– 5.288 *	+ 7.962 *	– 5.339 *	– 5.550 *

Table B 2-2 is continued …

Table B 2-2 continued ...

Explanatory variable	Car availability: Always	Car availability: Partially	National ticket ownership	Regional ticket ownership	Half-fare discount ticket ownership
Number of persons	1043	1043	1043	1043	1043
Number of observations	28808	28808	28808	28808	28808
ρ^2 (adjusted)	0.759	0.744	0.859	0.728	0.582

* Level of significance ≤ 0.10

Table B 2-3   Multinomial logit model for mobility tool ownership in groups not taking the panel effect into account (1985-2004)

Explanatory variable	No Car + No Tickets	No Car + HF T	No Car + Nat T / Reg T	Car + No Tickets	Car + HF T	Car + Nat T / Reg T
Age in years	− 0.038 *	− 0.028 *	− 0.163 *		− 0.015 *	− 0.090 *
Age in years squared	+ 0.001 *	+ 0.001 *	+ 0.002 *		+ 0.000 *	+ 0.001 *
Gender: Male	− 0.999 *	− 0.939 *	− 1.029 *		− 0.517 *	− 0.552 *
Nationality: Swiss national	− 0.691 *	+ 0.544 *	+ 0.183 *		+ 1.249 *	+ 0.818 *
College or university degree	− 0.095	+ 0.549 *	+ 0.700 *		+ 0.994 *	+ 0.956 *
In education	+ 0.225	+ 0.742 *	+ 0.866 *		+ 0.481 *	+ 1.036 *
Change in education	+ 0.205	+ 0.378 *	+ 0.320 *		+ 0.214 *	+ 0.236 *
Distance between the place of residence and the place of education in 1000 kilometres	+ 1.072	− 4.769	+ 1.985 *		+ 2.253 *	+ 2.355 *
In employment	− 0.887 *	− 0.195 *	− 0.088		+ 0.226 *	+ 0.371 *
Change in employment	− 0.233	− 0.047	+ 0.105		+ 0.105	+ 0.257 *
Distance between the place of residence and the place of employment in 1000 kilometres	+ 0.015	+ 0.012	− 1.395		− 0.144 *	− 0.261 *
Monthly income in 1000 CHF	− 0.088	− 0.075 *	− 0.069 *		+ 0.022	+ 0.058 *
Monthly income in 1000 CHF squared	− 0.037 *	− 0.006 *	+ 0.003 *		+ 0.003 *	− 0.001
Moving out of parents' house				+ 0.146		
Change in residence				− 0.040		
Birth of a person in the household				+ 0.236 *		
Number of persons in the household				− 0.020		
Number of rooms in the accommodation				+ 0.071 *		
Degree of urbanisation: Urban (referential category)						
Urban to rural				+ 0.728 *		
Rural				+ 0.920 *		
Place of residence abroad				+ 1.906 *		
Purchasing power index in the residential region				− 0.014 *		
Constant	− 0.318	− 1.954 *	+ 1.514 *		− 2.905 *	− 0.962 *

Number of observations	28808
ρ^2 (adjusted)	0.214

* Level of significance ≤ 0.10

Table B 2-4 Multinomial logit model for mobility tool ownership in groups taking the panel effect into account (1985-2004)

Explanatory variable	No Car + No Tickets	No Car + HF T	No Car + Nat T / Reg T	Car + No Tickets	Car + HF T	Car + Nat T / Reg T
Age in years	− 0.337 *	− 0.330 *	− 0.461 *		− 0.316 *	− 0.389 *
Age in years squared	+ 0.005 *	+ 0.005 *	+ 0.006 *		+ 0.005 *	+ 0.005 *
Gender: Male	− 3.331 *	− 3.331 *	− 3.468 *		− 2.963 *	− 3.018 *
Nationality: Swiss national	+ 0.148	+ 1.390	+ 1.008		+ 2.086 *	+ 1.637 *
College or university degree	+ 2.171 *	+ 2.836 *	+ 2.985 *		+ 3.313 *	+ 3.260 *
In education	+ 0.727	+ 1.219 *	+ 1.397 *		+ 1.034 *	+ 1.578 *
Change in education	+ 0.002	+ 0.194	+ 0.162		+ 0.075	+ 0.086
Distance between the place of residence and the place of education in 1000 kilometres	+ 0.846	− 4.055	+ 1.501		+ 1.824	+ 1.914
In employment	− 0.911 *	− 0.169	− 0.035		+ 0.300	+ 0.459
Change in employment	− 0.145	+ 0.068	+ 0.229 *		+ 0.229 *	+ 0.383 *
Distance between the place of residence and the place of employment in 1000 kilometres	+ 0.120	+ 0.112	− 1.601		− 0.033	− 0.261
Monthly income in 1000 CHF	+ 0.122	+ 0.069	+ 0.061		+ 0.117	+ 0.160
Monthly income in 1000 CHF squared	− 0.052	− 0.014	− 0.003		− 0.001	− 0.005
Moving out of parents' house				+ 0.335		
Change in residence				+ 0.020		
Birth of a person in the household				+ 0.629 *		
Number of persons in the household				+ 0.001		
Number of rooms in the accommodation				+ 0.012		
Degree of urbanisation: 　Urban (referential category) 　Urban to rural 　Rural				 + 0.406 + 1.000 *		
Place of residence abroad				+ 3.207 *		
Purchasing power index in the residential region				− 0.040 *		
Constant	+ 1.813	+ 0.321	+ 3.736 *		− 0.615	+ 1.311
Individual-specific random parameter	− 7.842 *	− 7.842 *	− 7.842 *		− 7.842 *	− 7.842 *
Number of observations						28808
ρ^2 (adjusted)						0.415

* Level of significance ≤ 0.10

Die VDM Verlagsservicegesellschaft sucht für wissenschaftliche Verlage abgeschlossene und herausragende

Dissertationen, Habilitationen, Diplomarbeiten, Master Theses, Magisterarbeiten usw.

für die kostenlose Publikation als Fachbuch.

Sie verfügen über eine Arbeit, die hohen inhaltlichen und formalen Ansprüchen genügt, und haben Interesse an einer honorarvergüteten Publikation?

Dann senden Sie bitte erste Informationen über sich und Ihre Arbeit per Email an *info@vdm-vsg.de*.

Sie erhalten kurzfristig unser Feedback!

VDM Verlagsservicegesellschaft mbH
Dudweiler Landstr. 99   Telefon +49 681 3720 174
D - 66123 Saarbrücken   Fax   +49 681 3720 1749

www.vdm-vsg.de

Die VDM Verlagsservicegesellschaft mbH vertritt

Printed by Books on Demand GmbH, Norderstedt / Germany